H. T. Hammel · P. F. Scholander

# Osmosis
# and Tensile Solvent

With 67 Figures

Springer-Verlag
Berlin Heidelberg New York 1976

Professor Dr. H. T. HAMMEL and Professor Dr. P. F. SCHOLANDER,
Scripps Institution of Oceanography, University of California,
San Diego, P.O.B. 1529, La Jolla, CA 92037/USA

The cover motif shows Fig. 29.

ISBN-13: 978-3-642-66341-3        e-ISBN-13: 978-3-642-66339-0
DOI: 10.1007/ 978-3-642-66339-0

Library of Congress Catologing in Publication Data Hammel, Harold T 1921—
Osmosis and Tensile Solvent Bibliography p Includes index 1 Osmosis 2 Solution
(Chemistry) 3 Surface tension 4 Water I Scholander Per Fredrik, 1905— joint
author II Title QD543 H253 76-3684

# Preface

This monograph has been written from our conviction that the present notions of the state of water in osmotic systems are obscure, if not incorrect. The basic ideas presented herein are for us not original, but they have previously been ignored. We shall attempt again to bring the essential concepts to the attention of the functional biologist with the hope that they will be duly considered and accepted. We even dare to expect that many will be able to recognize the inherent beauty in the old idea that all colligative properties of water stem exclusively from the fact that the water is under tension, regardless of the particular process by which it has been placed under tension in the biological system. The ideas are so simple and so obvious that we are amazed biologists have not already recognized their validity even without the assistance of physical chemistry and chemical thermodynamics, which deal with the subject of water under pressure in solutions and in matrices. We expect that drawing attention to experimental evidence heretofore not available will assist the more conservative physiologist to reconsider and reject fictitious notions about the properties of water in a solution.

The concept of osmotic processes expressed in the following essay, is the result of many pleasurable and adventurous years shared by the authors and a multitude of colleagues, attracted by the venerable old problem of how sap gets up into tall trees. Various experimental approaches took us from lofty northern trees into the muck of tropical mangrove swamps, and led to a series of experiments in various laboratories studying the effect of gravitational and magnetic forces on the osmotic process. Starting out as doubting Thomases on the old cohesion theory we were soon converted, but little did we anticipate the radical views it would lead to with respect to osmosis and imbibition. Neither did we know then that a beautiful, unifying view had already been expressed by the old masters around 1900-1910 and that this view - the same as ours - has been voiced independently by others several times up until quite recently.

During our off and on preoccupation with these
problems, we have relied heavily upon stimulating discussions and advice from a large number
of colleagues. In naming these friends it is,
of course, with the understanding that it im-
plies neither agreement nor disagreement with
our views.

We are greatly indebted to colleagues at this
University for discussions and advice and for
their laborious task of reviewing all or por-
tions of our manuscript: Professor Y. Fung,
AMES, UCSD School of Medicine; Professor G. Arrhenius, IPAPS, Scripps Institution of Oceanography; Professor A.B. Hastings, Neurosciences,
UCSD School of Medicine; Dr. E. Hemmingsen and
Dr. A. Yayanos, PRL, Scripps Institution of
Oceanography; Dr. A. Hargens, AMES, UCSD School
of Medicine; similarly Professor Frits Went,
University of Nevada, Professor L. Irving, University of Alaska and Professor J. Steen, University of Tromsø, Norway, and Dr. C.B. Wenger,
John B. Pierce Laboratory, have critically read
our manuscript.

We have had the great privilege of constructive
discussions with Professor H. Alfvén, AP&IS,
USCD and Professor W. Elsasser, University of
Maryland. Most rewarding were discussions with
earlier proponents of the tension theory of osmosis: Professor K. Herzfeld, Catholic University, Washington and Dr. K. Mysels, General
Atomic Company, San Diego.

One of us (S), on a Guggenheim Fellowship, had
the great privilege of spending half a year in
Stockholm in Professor H. Theorell's Lab, at the
Karolinska Institute and another half year in
Australia in Dr. H. Le Messurier's Lab in the
Medical School, University of Adelaide. These
were most rewarding visits and the latter
brought us in stimulating contact with Professor
R. Slatyer, CSIRO, Canberra.

In the Spring of 1973 Professor F. Went at the
University of Nevada invited the authors togeth-
er with Professor J. Dainty, University of
Toronto and Professor H. Currier, University of
California, Davis, to a three-day discussion on
osmosis in his wonderful hideout lab in Death
Valley, which proved most fruitful. Professor
R. Fernald and Professor D. Farner, University
of Washington, gave us the privilege of undis-
turbed and vibration-proof laboratory space at
Friday Harbour for exacting magnetic studies
on osmosis.

We wish to express our sincere gratitude to Miss Carole Mayo, who with exceptional expertise has handled the chores of all manuscript typing and other secretarial matters. Mrs. Marie Mathers has managed innumerable administrative services for which we are most appreciative.

Our work since its inception has been funded by numerous general or specific grants from the National Science Foundation, The National Institutes of Health, the Norwegian Research Council and the Rockefeller Foundation. Some of the work has been expeditionary involving Scripps Vessels, in particular the R/V Alpha Helix. We are greatly indebted to the excellent services by officers and crew in these adventures, which were always only a part of larger programs.

The photographs of the founders have kindly been furnished by: Dr. Gustaf Arrhenius, Scripps, UCSD: Mrs. Mireille Caunesil, Grenoble; Dr. D.W. Jeffrey, Trinity College, Dublin; Miss Sytska M. Klinkhamer-Mobach, Utrecht; Dr. Konrad F. Springer, Heidelberg; Dr. Frits Went, University of Nevada; Library of Congress; M.I.T. Historical Collections; Photo Harlingue-Viollet, Paris.

Although executed in the closest collaboration it will undoubtedly be perceived by the reader that Chapter I, dealing with an historical review and the experimental developments, is written by P.F. Scholander, whereas the remaining Chapters II-IV, dealing with the thermodynamic aspects, are the work of H.T. Hammel.


La Jolla, Spring 1976                    H.T. Hammel
                                         P.F. Scholander

# Contents

# *Chapter I* Perspectives on the Mechanism of Osmosis and Imbibition

## I. The Founders

The Pharaohs left to posterity one of the greatest wonders of human endeavor, epitomized in the pyramids. We read in "Daily Life in Ancient Egypt" by Waley-el dine Sameh (107):

> "What secret methods enabled the craftsmen of the Nile valley to work the great blocks of stone and the enormous statues of granite and diorite with tools of comparative soft bronze? Hewn blocks, (Fig. 1), weighing hundreds of tons were transported for miles from the mountains over desert sand, soft ploughed land and the Nile."

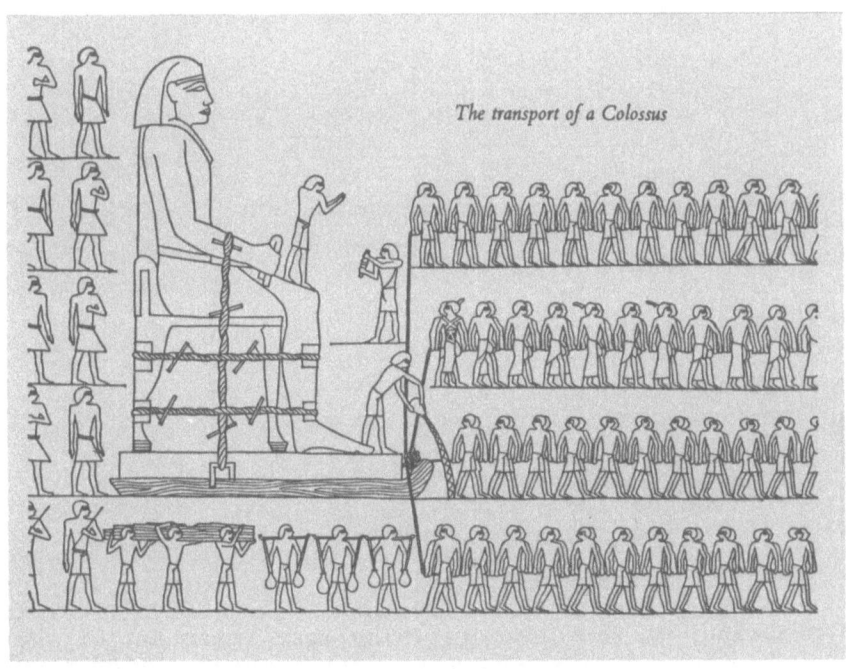

*The transport of a Colossus*

Fig. 1

> "...To detach a block of the required dimensions from its parent mass, a line was drawn along the limits and wedge shaped slots driven into it at regular intervals...Dry wooden wedges were driven into the slots and then soaked with water, particularly at night when evaporation was slight, till the expanding wood split off the block of stone..."

Now 4000 years later, we ask: What produces such a force when water is poured over dried wood? Is it a rise in the water pres-

sure as in an hydraulic jack? Hardly, for wood is leaky and a
leaky jack will not do. Would not a purely structural force
like the elastic pressure in a sponge be more likely?
Osmotic flow in one form or another has been observed since ancient times, and quantitative estimates appeared upon the scene
in the early 1800's, when Henri Dutrochet devised a quite sophisticated 'endosmometer' (Fig. 2.) He describes the endosmotic
phenomenon thus: "One observes that small animal bladders, filled with a dense solution and completely closed and plunged in
water swell excessively and become turgid" (25). Using pieces
of bladder or coecum as membrane and sugar solutions in the
osmometer he found that water entered at a rate rising with the
concentration, and that the pressure buildup, measured by a
mercury column, also rose with the concentration. He also noticed a faint rise with temperature on these parameters.

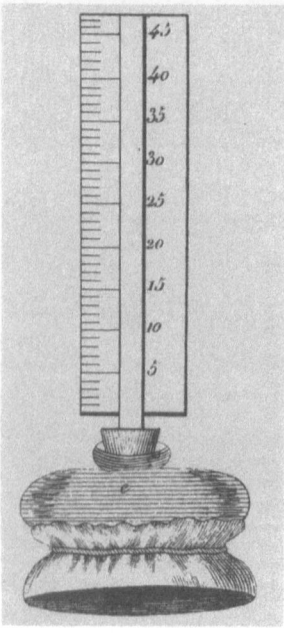

Fig. 2. An early 'endosmometer' by Dutrochet,
1837

The art of measurement took a big step forward when Moritz
Traube invented in 1867 (99) highly selective precipitation
membranes of various sorts. The most important of these proved
to be those employing copper ferrocyanide, often simply as a
plug at the end of a capillary. With this simple tool he measured weak osmotic pressures, or, as he saw it, "the attraction
of water by the solute". Ten years later came another breakthrough when, in 1877, botanist Wilhelm Pfeffer (62) conceived
the brilliant idea of precipitating Traube's membrane within
the porous wall of a ceramic tube, diffusing copper sulphate
from one side and potassium ferrocyanide from the other, or,
better, one after the other from the same side. This resulted
in a precipitate rigidly supported within the wall, which made
it possible for the first time to measure osmotic pressures
amounting to several atmospheres reproducibly in sugar solutions. Thus he accurately pinned down a close proportionality

to concentration and a linear, slight rise with temperature. One must admire the accuracy of his data, Figs. 3 and 4.

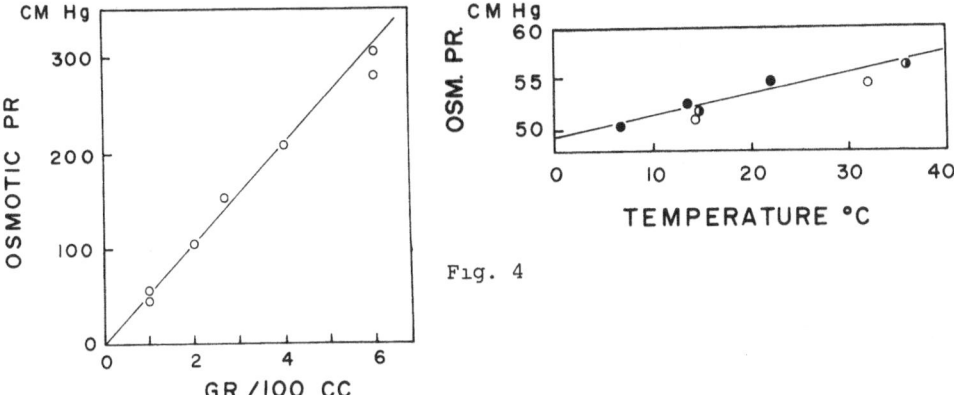

Fig. 3

Fig. 3. Osmotic pressures (π) of cane sugar solutions plotted from Pfeffer's Table 9, 1877. Temperature 14.8 + 1.3°C. The diagonal is the line derived from the van't Hoff gas analog

Fig. 4. Osmotic pressures (π) of three different 1% cane sugar solutions at different temperatures, plotted from Pfeffer's Table 12, 1877. The diagonal is the line derived from the van't Hoff gas analog

In 1878 the great French physical chemist François Raoult demonstrated in 18 different salt solutions a striking parallel in the lowering of freezing point and vapor pressure for each individual salt when compared with the pure water (67). The solutions were all made up from 4.5 g *anhydrous salts* dissolved in 100 ml water. The freezing point was determined in 100-120 ml solution, stirred and slowly cooled to the incipient freezing point plateau, and was readable to 1/200°. The vapor pressure lowering was determined by a differential manometer technique, ingeniously at 100°C, which is easy to maintain and where the vapor pressure is some 40 times higher than at room temperature. We have recalculated his tabulated data to give his observed values, which are entered on Fig. 5. The diagonal line is the relation to the gas analog unknown at that time, and demonstrates the excellency of the data.

Five years later Raoult published his law of the constancy of the molar freezing point depression in watery solutions of some 30 *organic* compounds (70). He states,

"In the course of these experiments, I was struck by a singular result; namely when one multiplied the freezing point lowering of 1 g dissolved in 100 g water with the molecular weight of the solute, one gets nearly the same figure."

We have again recalculated his tabulated data, comparing the inverse of the molecular weight with the observed freezing point depression (Fig. 6). All data fall remarkably close to the theoretical diagonal, which was not known at that time. He extended

Wilhelm Pfeffer
University of Bonn

François M. Raoult
University of Grenoble

Hugo DeVries
University of Amsterdam

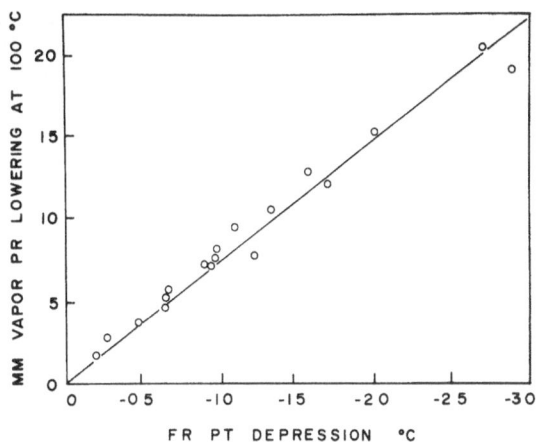

Fig. 5. Raoult's direct observations, 1878, on freezing point and vapor depression in 18 different salts. The diagram is recalculated from the original tables given in other units, and shows the close relation to van't Hoff's gas analog, represented by the diagonal, and the colligative aspects of these parameters. (-1.86°C corresponds to 13.7 mmHg at 100°C)

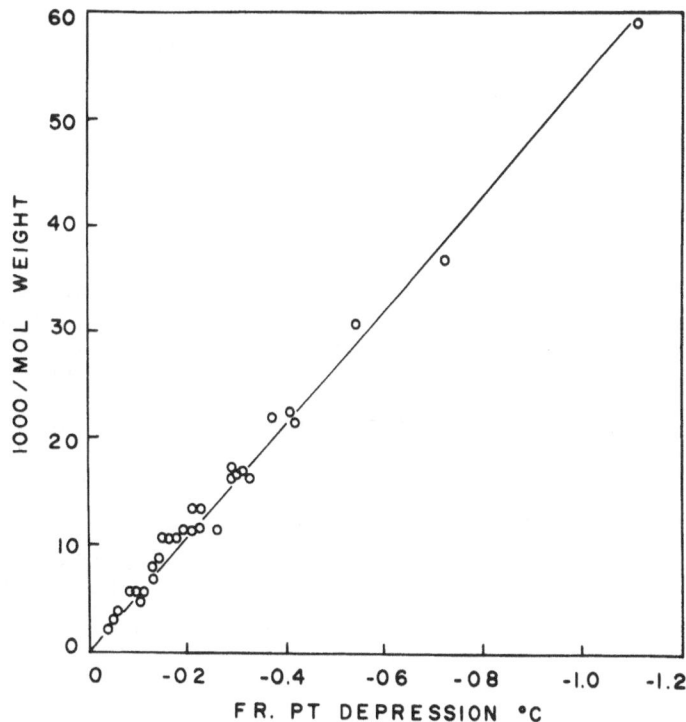

Fig. 6. Raoult's observations on freezing point depressions in 1% solutions of 30 organic compounds, recalculated from his Table, 1883, which is given in other units. The diagonal shows the excellent fit with van't Hoff's gas analog

vapor pressure and freezing point determinations of a variety of
organic solutes and solvents with similar accuracy, and established in numerous papers what we now call *the colligative
properties* of freezing point depression and vapor depression
(68, 69, 71).

At this time, 1877-1878, the University of Amsterdam added a
galaxy of stars to their science faculty, Johannes Van der
Waals, Hugo DeVries and Jacobus van't Hoff, respectively 43,
36 and 26 years of age. They had students in common and shared
an interest in the theory of solutions. Those most directly involved in osmosis were DeVries and van't Hoff.

DeVries, professor of botany, became intensely interested in osmotic phenomena in plants, and discovered in 1883-1888 that
equi-molar concentrations of a variety of organic compounds were
"isotonic" when tested microscopically by plasmolysis on selected cells from *Tradescantia (Rhoeo)* and a few other plants
(17-20).

He defined this in the following way: "Isotonic concentrations
are those, in which solutions of different solutes attract water with the same force." Taking incipient separation of the
plasmolemma as the point of isotonicity and solutions of potassium nitrate as a standard, he determined the relative osmotic
pressure of a great number of organic compounds and salts. Expressed on a molar basis his "isotonic coefficients" astounding-
ly lined up as whole number 2, 3, 4 and 5, and accurately depicted what was soon to be identified as the dissociation of
dilute electrolytes. We quote: (17)

> "Every acid and every metal have in all salts the same partial isotonic
> coefficient; the coefficient of the salts equals the sum of the partial
> coefficients of all of its components."

DeVries' coefficients in aqueous solutions of both salts and
organic compounds also closely mirrored the molar freezing point
depressions already found by Raoult, i.e. he perceived the colligative relation between these two parameters (18). His technique even yielded excellent figures for molecular weights (20).
One can only profoundly admire a genius who, with nothing but
living cells as a measuring instrument, so deeply penetrated
fundamental issues in physics and chemistry.[1]

Along similar lines H.J. Hamburger (1883, 33) discovered that
isotonic solutions could also be prepared for red cells, and
added the new observation that the osmotic balance remained
unchanged down to $0^{\circ}$C.

---

[1]DeVries is also famous for rediscovering Mendelean genetics, and is the
father of evolution by mutation, which he first recognized in an introduced
American evening-primrose *(Oenothera)*.

Svante Arrhenius                    Jacobus van't Hoff
University of Stockholm             University of Amsterdam

## Van't Hoff's Solute-Gas Analog

In May, 1884, a nova burst upon the northern sky when a student,
Svante Arrhenius, at the age of 24, submitted his doctoral thesis (2), "Recherches sur la conductibilité galvanique des électrolytes", to the University of Uppsala. It embodied in two volumes the essential ingredients for his celebrated dissociation
theory, which rested experimentally on electrical conductivity
measurements (Fig. 7) in solutions of known freezing point[2].

The stage was now set for one of the most brilliant generaliza-
tions in the history of physical chemistry, namely the solute-
gas analog, discovered by van't Hoff and embellished by Arrhenius. Besides a variety of strong intuitive leads, such as
Henry's law of dissolved gases, van't Hoff had several lines
of direct experimental evidence, such as: the isothermal proportionality between concentration and osmotic pressure (Pfeffer);
the linear temperature dependence of osmotic pressure (Pfeffer);
the constancy of molar vapor pressure lowering (Raoult); the
constancy of molar freezing point depression (Raoult); and the
isotonic series of solutions derived from plants and red cells
(DeVries and Hamburger).

---

[2]The fact that Arrhenius barely passed his dissertation so provoked the
great Ostwald that he journeyed to Uppsala to protect his young colleague,
and in 1903 Arrhenius won his Nobel laureate on the very thesis which had
been deemed unworthy of *jus docendi* (5, 76).

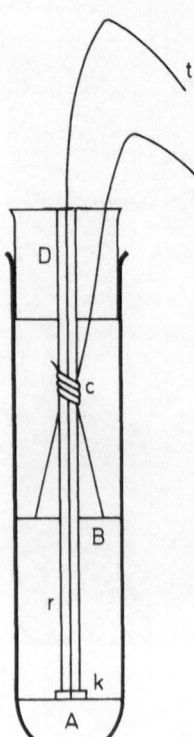

Fig. 7. Arrhenius' electrode assembly (1884) inserted in a test tube completely filled with the test solution. The rubber stopper $D$ is pierced by a glass capillary $r$. The electrodes $A$ and $B$ are circular platinum discs fitting snugly in the tube. The upper $B$ is firmly riveted to two platinum wires $c$, $u$, and the lower $A$ is sealed to the end of $r$ with a gasket and riveted to the platinum wire $t$. The unit is kept in an isothermal water bath

Upon a closer scrutiny of Pfeffer's osmotic data, van't Hoff realized that *the proportionality constant* was mirabile dictu, *the gas constant*, and that this was even reflected as a *proportionality with the Kelvin scale*. The ideal gas law $pV = nRT$ applied also to solutions! Thus the Boyle, Gay-Lussac, Avogadro, van't Hoff law was born. He further demonstrated its ready application to the law of mass action by Guldberg-Waage and to the simple temperature effect by Soret.

In 1886, van't Hoff broke his stunning news in a paper (103), "Une propriété générale de la matière diluée" in which he says:

"The pressure which a gas exerts at a given temperature, when a given number of molecules are distributed in a given volume, is equally great as the osmotic pressure, which under the same conditions would be produced by "most solutes", (our quote) when they are dissolved in an arbitrary solvent."

On reading this, Arrhenius (3) saw in a flash that

"the van't Hoff law was valid not only for "most solutes", but for all bodies, including those that before had been considered exceptions, namely electrolytes in a watery solution."

Another of van't Hoff's papers in December 1886 (104) dealt further with the troublesome electrolytes. Arrhenius vividly recalls his reaction to this paper in his autobiography (see Riesenfeld (76)):

"I devoured it at once...It was immediately clear to me that the deviation of electrolytes in watery solution from the van't Hoff-Raoult's law, regarding freezing point depression, embodied the strongest proof of their disintegration into ions. I had now two different ways of calculating the degree of dissociation, one, with the aid of freezing point depression, the other through the conductivity. In the overwhelming number of cases they gave the same result, and I could now openly pronounce the dissociation of electrolytes. One may imagine my great delight." (cf. Fig. 8).

Discussing the aberrations from the Avogadro's analog in his major paper of 1887 (105), van't Hoff pays full tribute to his young colleague:

"It must accordingly be admitted that the deviations of this kind in solutions are much more common, and show up in compounds where a cleavage in the ordinary sense is hard to perceive; to watery solutions belong thus most salts and strong acids and bases, and indeed the existence of the so-called normal molecular freezing depression and vapor lowering was first discovered when Raoult turned his attention to organic compounds; for then the normal behavior was practically without exception. It seems accordingly worth the risk to advance an Avogadro's law for solutions such as here done, and I would not have decided to do so, had not Arrhenius briefly pointed out to me the likelihood that the essence regarding salts and the like was a splitting into ions."

As Avogadro generalized Dalton's law by *associating atoms*, so Arrhenius generalized van't Hoff-Raoult's law by *dissociating molecules*. Let us briefly look at the evidence.

In Fig. 6 we presented the freezing point depressions of 30 different organic compounds observed by Raoult in 1% solutions. From four additional papers by the same author, Arrhenius picked a series of compounds for which electrical conductivity measurements had similarly been made by Kohlrausch (for references see (3)). These comprised 12 non-conducting organic compounds, 37 salts, 15 bases and 23 acids, all with widely

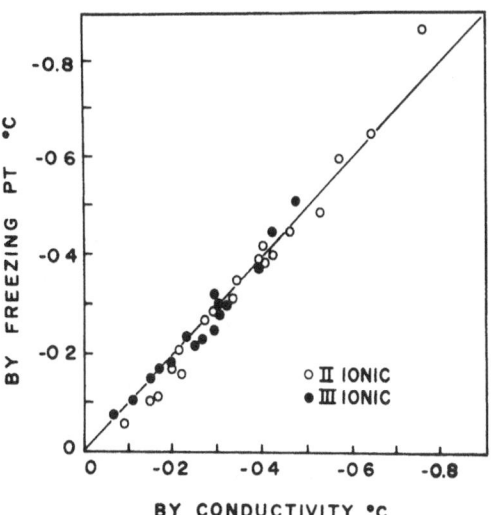

Fig. 8. Freezing points in 1% salt solutions by direct measurements (Raoult), versus calculations by Arrhenius from Kohlrausch's conductivity measurements. Our plot is recalculated from Arrhenius' Table, 1887. The diagonal is the line of equivalence

scattered freezing points and conductivities. Arrhenius' idea was that the degree of dissociation would be reflected by the observed *molar conductivity*, divided by the maximum this would attain upon further dilution. In turn, this would give the number of osmotic units and thereby the freezing point:

"When this activity coefficient α is known, one may calculate the value $i$ in van't Hoff's Table, i.e. the relation between the observed osmotic pressure and that which would result from exclusively inactive, (non dissociated) molecules... When $m$ is the number of inactive and $n$ the number of active molecules, and $k$ the number of ions from active molecules (e.g. for KCl $k = 2$, namely $K^+$ and $Cl^-$, for $BaCl_2$ and $K_2SO_4$ $k = 3$, namely $Ba^{2+}$, $Cl^-$ and $Cl^-$ resp. $K^+$, $K^+$ and $SO_4^{2-}$), then we have $i = \dfrac{m + kn}{m + n}$.

As the activity coefficient (α) is $\dfrac{n}{m+n}$ it follows that $i = 1+(k-1)\alpha$.

... On the other hand one may, according to van't Hoff, calculate $i$ from Raoult's freezing point determinations in the following way. The freezing point depression $t$ produced by dissolving 1 grammolecule in 100 ml water is divided by 18.5."

This data comprises the last column in Arrhenius' Table (1887) and pertains to 1 g solute per 100 ml water (not per liter as stated).

In Fig. 8 we presented the observed freezing points of 37 salts (Raoult) compared to those calculated from Arrhenius' activity coefficients: the match is excellent. In Fig. 9, we compare both observed and calculated freezing points with the inverse of the

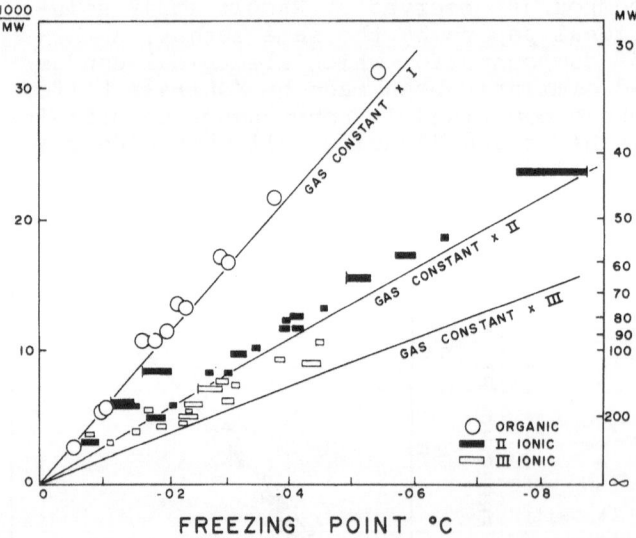

FREEZING POINT °C

**Fig. 9.** Inverse molecular weight (1000/*MW*) versus freezing point in 1% water solutions of organic compounds and salts. The length of the bars is the spread between direct measurements and values based on Arrhenius' dissociation constants. The vertical mark on the long bars marks the direct measurement. Our plot is recalculated from Arrhenius' Table, 1887

molecular weights and find excellent agreement with the gas analog for *organic* compounds. It will be noticed that for the ionic
compounds the direct and indirect values are also mostly very
close; but the *degree of dissociation* (i.e. distance to ideal
line) is variable and depends upon the nature of the salt, as
is well known. Not plotted in the diagram are the bases and
acids; the former behaved very much like the salts, the latter
showed considerable spread.


## Van't Hoff on the Mechanism of Osmosis

In the major paper referred to above, "The Role of osmotic pressure in the analogy between solutions and gases", van't Hoff
(105) states (cf. Fig. 10):

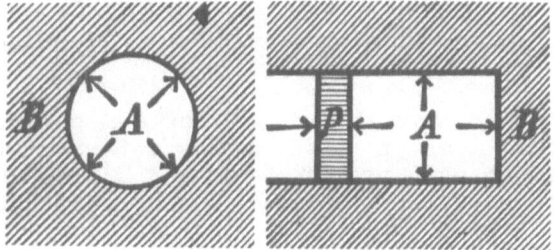

Fig. 10. *A* solution. *B* water.
*P* piston. (After van't Hoff's
originals, Fig. 1 left, Fig. 2
right)

"Let us introduce from a theoretical point of view the entity which shall
be the main subject of the following discussion. Consider a vessel A, submerged in water B and completely filled with a sugar solution (Fig. 1).
The wall of the vessel is completely rigid; it is pervious to water but
not to dissolved sugar. As is well known, *the water-attracting property*
(our italics) of the solution causes an influx of water into A. The amount
is very small, and the resulting pressure soon reaches a limit. We have
now reached equilibrium, and we have a pressure on the wall of the vessel
which we shall call the osmotic pressure."

"Obviously, we may also establish this equilibrium in advance, i.e. without previous water entrance, by connecting A with a piston, which applies
a pressure equal to the osmotic pressure (Fig. 2). One realizes then that
one may change the concentration of the solution simply by changing the
piston pressure, namely, by moving the water through the vessel wall in
one or the other direction."

He then outlines an osmotic experiment such as performed by
Pfeffer using a copper ferrocyanide membrane:

"These unusual membranes which we shall call "semipermeable walls" have
been extensively used also when not altogether perfect. They allow us
to treat solutions in a way very much like gases. This follows when we
consider that the pressure so characteristic for gases is now introduced
as osmotic pressure of solutions. We must emphasize that we are dealing
not with a forced artificial analog but with a basic one; for the mechanism which according to our present idea produces the pressure of gases,
and the osmotic pressure in the solutions is essentially the same: in the

first case we deal with the impact of gas molecules on the walls of the container, in the second with *the impact of solute molecules on the semi-permeable membrane*, (our italics) for the permeating solvent is on both sides and hence is out of the question."

He here perceived one basic ingredient of the osmotic mechanism, namely the effect of the thermal impact of the solute on a membrane, but this alone gives no hint of the full story and his "solute bombardment" theory soon fell into discredit. However, in almost the same breath, he offered the alternate common view: that the rise of pressure in the osmotic cell is caused by entrance of water due to "water attraction".

"...if one considers the osmotic pressure to be of kinetic origin, i.e. caused by bombardment from solute molecules, then the proof of this would entail proportionality of hits per unit time with the number of hitting molecules per unit of space. The proof would then be exactly the same as that pertaining to Boyle's law for ideal gases. If, in contrast, we regard the osmotic pressure as an expression of a water-attractive force, then its magnitude would obviously be proportional to the number of molecules per unit of volume..."

Let us here anticipate our own position by stating that the pitfall lies in van't Hoff's interpretation of the correlation between entrance of water and buildup of osmotic pressure: entrance of water is a *consequence* of solute bombardment at the boundaries of the solution; it does not *cause* the pressure but is a condition for it. Van't Hoff's illusion that the solution attracts the water, a small amount enters, and the resulting water pressure soon reaches a limit, has dominated the issue ever since.


## II. Thermodynamic Laissez Faire

Without empirical evidence for either "solute bombardment" or "water attraction" van't Hoff had reached an impasse. This is evident from a famous statement he wrote against a contender in 1892 (106):

"Again we have the basically pointless question: What exerts osmotic pressure? Really, as already emphasized, I am concerned in the end only with its magnitude; since it has proved to be equal to the gas pressure one tends to think that it comes about by a similar mechanism as with gases. Let he, however, who is led down the false path by this, rather quit worrying about the mechanism;"

His contender was no other than Lothar Meyer (53), who had shared with Mendeléev the Davy medal of the Royal Society. Van't Hoff continues that Meyer "is getting himself into grave danger" and shall "remain forever bewildered over his conclusions." In fact, such a dire foreboding coming from van't Hoff himself clamped the lid on the issue for a long time to come.

In his monograph on osmotic pressure in 1913, Alexander Findlay
(27) states a commonly voiced opinion:

"A solution does not, in itself, have any osmotic pressure; and the term
is used, in a somewhat loose manner certainly, to denote the hydrostatic
or mechanical pressure *which would be produced* if the solution were separated from the pure solvent by a membrane or septum which is permeable
only to the solvent."

*We* shall emphasize that the osmotic pressure, *alias* the dispersal pressure of the solute, is always there.

As to mechanism Findlay echoes:

"On the one hand, there is the kinetic interpretation of osmotic pressure,
according to which the pressure is due to the bombardment of the semipermeable membrane by the solute molecules, in a manner analogous to the
kinetic explanation of gaseous pressure: and on the other hand there is
the view that osmotic pressure is a hydrostatic pressure produced by the
entrance of solvent into the solution."

His concern, however, is mainly thermodynamic, about which he
states:

"Valuable as the thermodynamic theory of solutions has proved to be it
throws but little light on the mechanism of osmosis or on the intimate
structure of solutions. Nor has it ever claimed to do so."

In "Elements of Physical Chemistry", 1960, S. Glasstone and
D. Lewis (29) put into a nutshell the motivation for our present work:

"In recent years there has been a growing opinion, first voiced by van't
Hoff himself in 1892, that the actual mechanism of osmotic pressure is
not important. Both solvent and solute molecules will undoubtedly bombard
the semipermeable membrane, and distillation may also occur at the same
time through the pores. All this is, however, immaterial; in the study
of osmotic pressure the essential point is that, *for some reason connected with the presence of solute molecules*, the "escaping tendency" or
"activity" of the solvent molecules is less in a solution than it is in
the pure liquid."

Like Findlay they emphasize:

"The osmotic pressure is brought into existence *only* when the solution
is separated from the solvent by a semipermeable membrane. The resulting
osmosis, or tendency for osmosis to occur, then produces an excess pressure in the solution."

From the great textbook "Thermodynamics" by G.N. Lewis and
M.R. Randall, 1961 (44) we learn,

"The formal similarity...to the perfect gas equation has led to attempts
to interpret osmotic pressure in terms analogous to the kinetic theory
of gases. We do not believe that consideration of the collisions of solute
molecules with the membrane constitutes a useful approach to an understanding of osmotic pressure. Rather one should note the tendency of any

"gas to fill its container and of any solute to diffuse throughout the solvent accessible to it. The driving force in each case is the entropy, i.e. the increase in probability which arises for distribution of particles through a large volume as compared with a small volume. We expect the probability relationship to be the same in each case."

## III. Some Fundamental Experimental Facts

Amongst the plethora of more or less plausible mechanisms suggested to explain osmosis we have mentioned solute bombardment and solvent entrance. Others invoke solute-solvent affinity, surface tension, vapor gaps, electrostatic forces, entropy of mixing, etc. (27), and what is real and what is fiction can be resolved by new experimental fixes.

Current ideas on the mechanism of osmosis run in two main streams which are mutually exclusive; on the one hand is the generally accepted theory, which we shall call the "water concentration theory", championed by Mauro (48), Ray (72), Dainty (16), Slatyer (93), and we find it in many textbooks. On the other hand, we have the various aspects of the "water tension theory" which have been independently developed by Hulett (41), Herzfeld (40), Mysels (55), Duclaux (23), and ourselves (35b, 36, 77, 78, 80-82, 84, 89, 90).

It is essential for the following discussions to keep in mind some old and some relatively recent experimental facts pertaining to any rational explanation of the mechanism of osmosis.

### 1. Lowered Vapor Pressure Over Solutions and Unsaturated Gels

### 2. Osmosis by Bulk Flow

Impetus in much recent activity lies in the growing realization that there is a strong component of hydraulic flow in many instances of biological osmotic transport. Thus, it has been found that osmotic flux is frequently many times greater than diffusive permeability determined by tracer experiments (Ussing, 1952, 102) and it has been calculated that when the pore sizes exceed some 2 nm the hydraulic flux becomes overwhelming (Pappenheimer, 1953, 59). Experimentally Mauro (1957, 48), and Meschia and Setnikar (1958, 52) found that the flow of water through a collodion membrane into a dextran solution could be as much as 600-700 times faster than self diffusion, as determined by labeled water. Even allowing for a doubling of the effective membrane thickness by an unstirred layer the flow rate would still be vastly faster than diffusion and would demonstrate hydraulic flux, i.e. bulk flow. The fundamental conclusion is that osmotic flux through a membrane may be either entirely diffusive, or *overwhelmingly hydraulic*, or it may be mixed.

## 3. Equivalence of Hydraulic and Osmotic Flux

In their brilliant and exhaustive studies of osmotic pressure
the Earl of Berkeley and E.G. Hartley, even in 1909 (10), found
that water passed through a copper ferrocyanide membrane at the
same rate when activated by many atmospheres osmotic pull of a
stirred sugar solution as when activated by an equally steep
gradient in hydrostatic pressure of pure water. Mauro (1957, 48)
and Meschia and Setnikar (1958, 52) confirmed this important
point when working with dextran solutions and collodion membranes. Hargens (1972, 37), using a stirred dextran solution
with an osmotic pull of 40 cm of water and a dialyzing membrane,
found the osmotic and hydraulic rates to be within ±2% of each
other.

## 4. Osmotic Flux Moving against the Water Potential

In careful experiments on *hydraulic* osmosis, Meschia and
Setnikar (52) opposed a stirred dextran solution against a
90-times stronger urea solution across a collodion membrane.
The osmotic flux went against the water potential, i.e. from
the urea into the dextran. Similarly Hargens (1972, 37) with
a dialyzing membrane, found that a stirred 0.005 mol dextran
solution drew in a 0.2 mol NaCl solution, approximately 800
times stronger, at a rate which was within 5-10% of that of
pure water.

# IV. Water Tension Theory

Ever since the brilliant era culminating with van't Hoff, we
have been left with the basic question: what is the nature of
the cause of the colligative properties and why does the osmotic
pressure of ideal solutions faithfully reflect the kinetic state
of gases? The *kind of solute* does not matter, so the question
narrows down to: what physical property of the *solvent* changes
this way in the presence of the solutes? Escaping tendency,
activity, and fugacity are merely descriptive terms. We may
come back to the same question we raised about the swelling
wood (p.1): When a semipermeable bag filled with a solution
is placed in water the bag swells under pressure; is the water
under pressure then, like the fluid in an hydraulic jack? How
could that be when the barrier is pervious to water? We shall
endeavor to show that all of this resolves itself into the reality of negative solvent pressure resulting from elastic expansion of matrices and thermal dispersal pressure of solutes.

The idea of solvent tension was already perceived by Arthur
Noyes (1900, 56), and was expounded with great clarity and com-
pleteness by George Hulett (1902, 41). Since then it has surfaced independently several times, such as in 1937 by Karl
Herzfeld (40), by Karol Mysels in 1959 (55), by Jacques Duclaux
in 1965 (23), and ourselves in 1965 (84).

## Physical Models of Imbibition and Osmosis

Let us start by scrutinizing the kind of systems that display osmotic features, to determine exactly their criteria or symptoms. In the most general terms, we may say that osmosis and imbibition constitute the science of water relations in dispersed systems, of which two classes may be discerned. One has discrete particles that may range in size: from sand through clay, starch, colloids, microsolutes to ionized salts. If suspended in water and orbited at zero gravity in a space craft all of them (even sand) would disperse evenly in the water by thermal motion, i.e. by kinetic action. The other class comprises nonkinetic matrices where water is held in a capillary or diffuse meshwork, such as xylem, the cell wall of plants, porous ceramics and gels. It will be seen immediately that *sedimented* sand, or starch, passes imperceptibly into matrices, as do likewise high concentrations of macromolecules when they *crowd*, or *pack*, together. Because of the gradual transitions between these states there is a strong *a priori* reason to expect that their water relations would show such basic similarities as to allow imbibition and osmosis to be combined under a common term: osmotic processes (81).

Let us clarify the main points by two examples, pictured in Fig. 11. A is a block of *moist* gelatin placed in a dish of water. B is a *slack* dialyzing bag filled with a dextran solution, also placed in water. The two systems display five identical responses to water:

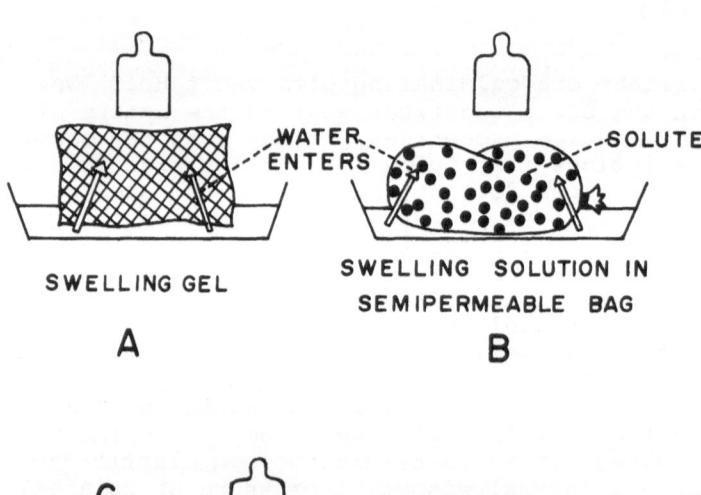

WATER ENTERS     SOLUTE

SWELLING GEL     SWELLING SOLUTION IN SEMIPERMEABLE BAG

A     B

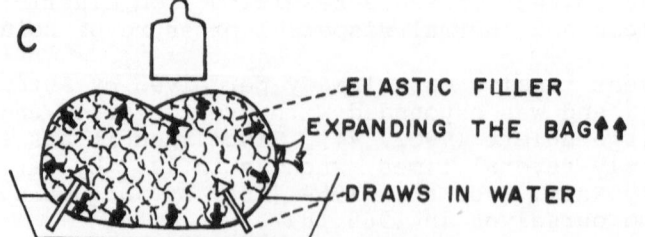

ELASTIC FILLER
EXPANDING THE BAG⬆⬆
DRAWS IN WATER

Fig. 11. Schematic presentation of negative pressure theory for imbibition and osmosis

1. Both have initially a lowered vapor pressure.
2. Both swell when put in water.
3. Both can lift a weight.
4. When the swelling becomes balanced by the weight, the vapor pressure returns to that of ambient water.
5. If a permeant solution is used in the dish instead of water, the solution is drawn into the gel or bag irrespective of the water potential.

Our questions are: what is the mechanism behind this strikingly similar behavior? What causes the osmotic pressure? What causes the swelling pressure?

Is there not, one may ask, a simple physical model which behaves as do our solution and gel? Indeed there is (Fig. 11C). Stuff a dialyzing bag with a springy substance (such as a shredded rubber sponge), flood it with water, squeeze out half of the water; with the pressure still on, close the bag, and there you are! The compressed filler puts the water in the slack bag under tension, which in turn lowers the vapor pressure. Water is *drawn in* from the dish, and the bag swells. The compressed sponge (spring) will lift a weight. When the spring can lift no more, the water pressure becomes ambient, and the vapor pressure returns to normal. If the bag is placed in a *permeant* solution of any strength, the negative pressure in the bag draws it in. Here then are all the points of similarity displayed in common with the solution and the gel!

We shall at once notice one cardinal point: that the water in the bag is initially *under tension*, i.e. below ambient pressure. The outside water is hence *drawn* into the bag, running from ambient to a negative pressure, *down* a pressure gradient. The *influx* of water does not lift the weight, but *it permits the spring* to do so which resolves van't Hoff's paradox (p. 12).

The next question is: can we identify a "spring"-like mechanism tending to expand the boundaries of a solution and a gel? The answer is yes, indeed; the kinetic (thermal) dispersal pressure of solute molecules and the (mechanical) elasticity of the matrix in a gel! As the compressibility of air has some resemblance to Hooke's[3] law of the spring, Robert Boyle referred to his gas law as the "spring of the atmosphere". Similarly one may refer to the dispersal pressure of solute molecules as the "spring of the solutes". The "spring of the matrix" is its elasticity, measurable by compressing it under water.

We shall describe the measurement of these elastic, spring-like driving forces and demonstrate their effect on the water, and shall identify a) the negative pressure and b) the spring.

---

[3]Robert Hooke was a close associate of Boyle at Oxford.

## Cohesiveness of Water

It sounds preposterous to claim that water in practically all
of its most intimate relations with our own body is under high
negative pressure: in our blood, sweat and tears, not to speak
of epicurean delights! And yet, this is the heart of our thesis.

The fact that fluids can sustain considerable tension was demonstrated by Marcellin Berthelot in 1850 (12). By heating a sealed
glass tubing containing water and a small gas bubble, he dissolved the bubble. On cooling the glass he found that the water
suddenly ruptured and from the temperature and thermal expansion
coefficients of glass and water he calculated the tension as
-50 atm for water. It was twice as much for alcohol. Many experiments later confirmed and expanded his finding, the most
direct being those of Lyman Briggs in 1950 (14). A Z-bent capillary was filled with water and spun in a centrifuge (Fig. 12).

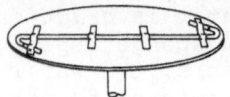

Fig. 12. Simplified version of Brigg's method for demonstrating negative fluid pressure. Water filled capillary
taped on thin aluminum rotor (80)

The centrifugal force (tension) at the center was calculated
from the rotational speed and the radius to the menisci. He
found the maximum tension before rupture of the water to be
about -270 atm at $10^\circ$C. The cavitation, instead of taking place
within the water body, is most likely caused by nucleation sites
at the wall interface. Various ways of calculating the theoretical cohesiveness within water have yielded figures usually between -1000 to -2000 atm (Yayanos, 1970 (109); Oertli, 1970 (57);
Apfel, 1972 (1). The highest experimental value may well be that
reported by Roedder, 1967 (76b). In a modern Berthelot-kind of
experiment he slowly froze a mineral containing trapped water
with an air bubble. The expanding ice caused the bubble to dis-
solve in unfrozen water, and upon warming the temperature of
melting ($+6.5^\circ$C) was obtained. The cohesiveness was estimated
at about -1000 atm.

A familiar case of sub-ambient pressure is ordinary *capillarity*,
such as in a glass tube or a filter paper. We observe that a
capillary standing in a dish pulls up the water. This is caused
by glass-water forces operating at the meniscus and is therefore
inversely proportional to the circumference. It reaches 30 m
at a 1 micron diameter, i.e. at the meniscus there is a negative
pressure of 3 atm compared to the flat surface in the dish. When
a capillary is held in air and contains a short column of water,
symmetrical pulls from the menisci put the water under tension.
This *tension*, of course, must be balanced by an equal and opposite elastic *compression* of the glass (82).

Analogous to the glass capillary is any wet matrix, such as clay,
wood and paper. In drying, menisci are formed at the surface,
and as evaporation progresses the matrix becomes increasingly

compressed, stretching the water into tension. The maximum negative pressure is inversely related to the diameters of the largest pores. When these are very fine, of $10^{-1}$ nm dimensions, the
compression becomes prodigious. The vapor pressure at $20°C$ is
17.5 mm Hg. At 90% humidity it would be 1.75 mm less. According
to Poynting's differential relation (p. 43) this would correspond to some -140 atm hydrostatic pressure: i.e. a gel would
be compressed by 140 atm, and will swell by this force when re-
moistened. At lower humidity, which is much more prevalent, many
hundred atmospheres compression may be obtained, as measured by
Lloyd and Moran (45). This applies to porous materials like
leather, clay, wood (p. 1), etc., lying around in nature. The
potential energy built up by evaporative compression, effects
the "winding of the spring", which is released when water is
admitted.

If we further realize that the thermal motion of *solute molecules*
also strain against the free surface we perceive how common is
negative hydrostatic pressure in water. Indeed, the dispersal
pressure of the salts in the sea water amounts to some 25 atm,
and lowers by the surface effect the hydrostatic pressure of
the pure water solvent by 25 atm in the entire ocean!

## A. Matrices

### Measurements of Tension and the Role of the Free Surface

How then can we measure these negative fluid pressures? Let us
first create a simple system with obvious negative pressure
(Fig. 13A). We shall use a syringe in which is inserted a compressed spring which pushes the plunger outwards with a constant
force $p$. The syringe is filled with water and a plug of cotton
put in the nozzle. The capillarity of the plug prevents air from
being sucked in, and with the system in air we have the water
under permanent negative pressure $(-p)$.

There are two obvious ways of measuring the negative fluid pressure in our syringe. One method is to push on the plunger
(Fig. 13B), until the water menisci uncouple from the cotton
plug. The force $p$ on the plunger is then that of the spring
which causes the negative pressure. This *push method* is applicable to any tension. In the other method (Fig. 13C), the *pull
method*, suction is applied to the nozzle until the water is
free. It can be used only for weak tensions, at most a few me-
ters of water. Both techniques measure the same thing, namely
*the matrix (spring) expansion*, which produces the tension.

The pull method may be operated simply by inserting a cotton
wick in one end of a 1-2 mm ID flexible plastic tubing; the
other end is connected to a glass capillary and a manometer de-
vice; or frequently (Fig. 14) one may simply measure the distance $H$ of the capillary below the wick when the fluid is in
balance, adding the capillary blank $C$, to produce the water
tension.

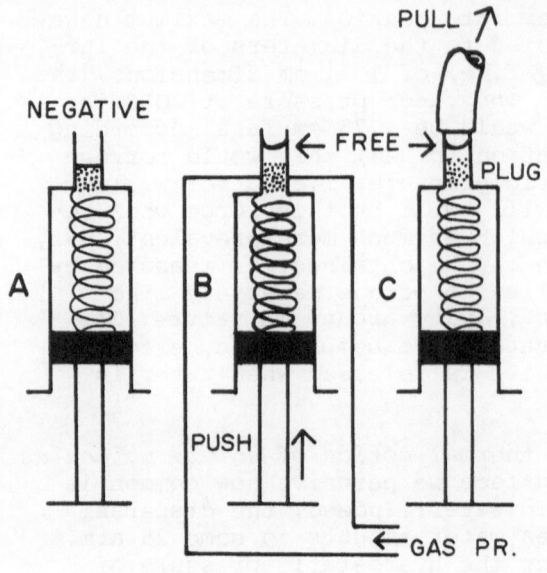

Fig. 13. Weight and frictionless syringe furnished with expansion spring in barrel and porous plug in nozzle. *A* water under negative pressure which is measured by *push* (*B*) and *pull* (*C*), until meniscus in nozzle is free. Either method measures the compression of the spring which causes the tension in the water (81). The push method *B* is usually performed by enclosing the system in a pressure chamber except for the nozzle or its analog

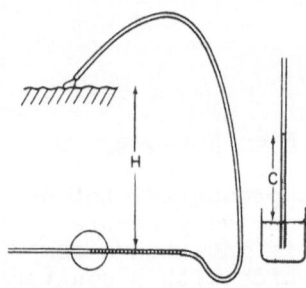

Fig. 14. Measurement of weak fluid tensions by means of wick probe. The negative pressure equals $H + C$, $C$ being the capillary blank (80)

With this simple tool it is easy to show that water under tension is found everywhere, in moist sand, wood, paper, leather, gels, dough, clay or any pasty substance. Foams, like whipped cream and shaving lather have a few centimeters of negative fluid pressure. The matrix here consists of fine air bubbles crowded together by a scant fluid volume (81).

A simple demonstration of the role of the free surface is presented in Fig. 15. A wick was placed on sedimented starch in a shallow tray and connected with a manometer. The tray was suspended from a torsion balance which measured the evaporation loss. The pressure remained ambient until the receding surface contacted the starch. At this moment the surface turned opaque and negative pressure developed sharply as menisci curved down, packing the grains. When water is added, the matrix relaxes (swells). *There is no effect on the water until the matrix strains against the free surface.*

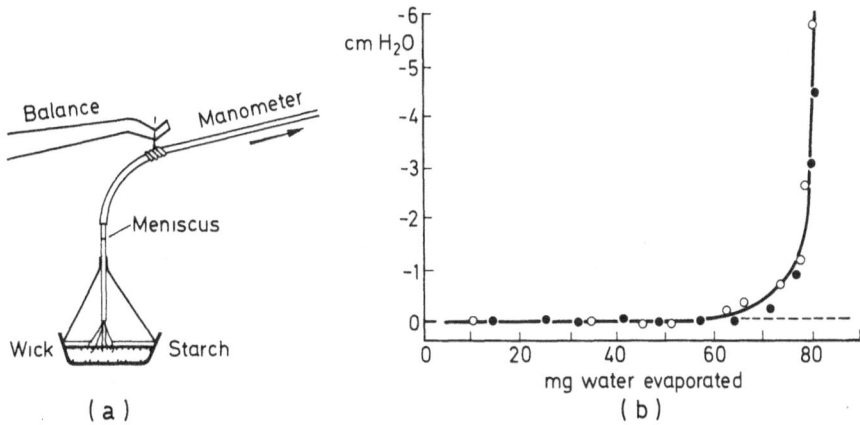

(a)  (b)

Fig. 15 a and b. (a) Simultaneous measurement of evaporative water loss and tension in sedimented starch, being touched by a wick. (b) Water tension versus water loss in sedimented corn- and potato starch. The sharp rise occurs when water surface touches sedimented grains (81)

## Gravitational Pressure Gradients

Let us now take a closer look at the water-matrix interaction in a starch suspension (Fig. 16). In the left column the starch is sedimented on the bottom, leaving the surface free. If we thrust a rod into the bottom layer and try to stir, we find it very stiff (thixotropic), and the layer will fracture before it will yield to quick strokes of the rod. In other words, there is a strong interaction between the starch grains themselves. It is now of interest to know whether there is also a *starch-water* interaction in the system which would modify the gravitational

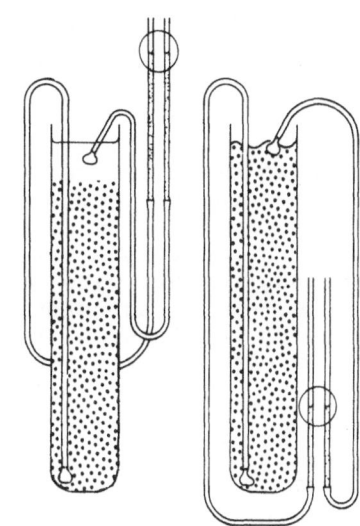

Fig. 16. Wick measurement of hydrostatic gradient in sedimented starch columns. *Left*, there is a free surface and both capillaries maintain their menisci at the blank value. *Right*, starch bucks the surface and lowers the water pressure to negative (subambient) values; in each case the two capillary readings are identical (82)

pressure gradient in the water. By putting a wick probe at the
top and one at the bottom we can compare the pressure readings
in the pure water, taking care to use clean and identical capillaries. It will be seen at once that the readings in the pure
water from surface and bottom *must* stand level, irrespective of
the starch. If they did not we could connect the capillaries
and create a perpetual flow.

When we add more starch it finally crowds up against the surface, which becomes opaque and bumpy. As the starch compresses,
the water pressure becomes *negative* and the consistency pasty.
What about the gravitational pressure gradient? Again, if the
depressed menisci in the capillaries did not stand level the
second law of thermodynamics would be violated. Empirically
(81, 82) there is no interaction with the mobile water nor can
there be in systems of this sort. *The water is not diluted by
the matrix, only interrupted. The hydrostatic pressure is transmitted undiminished in all directions in accord with the principle of Pascal.*

## Negative Pressure by Crowding and Conformational Changes

The role of the free surface, or simply of capillarity, in developing negative pressure is strikingly demonstrated in three
sedimented systems described below:

1. The colloidal osmotic pressure is determined in a blood
sample using an osmometer with a dialyzing membrane and charged
with isotonic saline on the other side. The sample is removed,
centrifuged and the plasma is quickly decanted. Transferred
back to the osmometer the compacted cell mass shows a negative
pressure several times higher than did the blood. The reason is
that the red cells were packed and compressed in the centrifuge,
and afterwards with the supernatant plasma quickly removed they
expanded and strained against the surface of the small volume
of the remaining plasma. This large "crowding pressure" of the
cells is simply added to the colloidal plasma pressure already
present (81).

2. Another case of the effect of crowding upon negative pressure
may be illustrated by *conformational* changes in suspended materials. Sephadex C50 consists of a powder of microscopic spheres
of insoluble carbohydrate. When it is poured into the water the
spheres swell and sediment. If more powder is added it finally
reaches the surface and packs, forming a paste with negative
water pressure, as measured by the wick method (p. 21). When a
trace of salt is stirred into the paste each sephadex granule
shrinks to half size or less, and the paste returns to a fluid
state, as the shrunken spheres leave the surface and sink, re-
lieving the tension. Here is a change in the water pressure that
is purely conformational in origin, involving no change in numbers of particles, only in their size.

3. Finally, if salt is added to a pasty suspension of bentonite
clay it stiffens, which is to say that the original negative
pressure becomes exaggerated by the increased space requirement
per particle, electrostatic or otherwise. One may predict that

purely conformational changes in macro molecules involving their
size, such as oxygenation of hemoglobin, will have an osmotic
read-out in concentrated solutions, by changing the crowding
pressure (81).


## B. Matrix-Solute Balance

This Chapter is not a general discourse of water relations in
plants, nor on the extensive recent literature dealing with water potential derived from vapor pressure measurements (cf.
Slatyer, 93). Rather, we shall deal only with direct hydrostatic
evidence for tensile water in osmotic systems, for this, in our
view, is the sole causative factor in all colligative properties (35b, 36).


## Sap Rising in Trees and Vines

The spectacular matrices which constitute the xylem of trees
and vines have fascinated not only biologists, but physicists
as well, for over a century. Certain conifers like redwood and
Douglas fir, as well as one species of eucalyptus rise to nearly 100 m, or higher, and various vines like grape, wistaria
and rattan palms also grow to heights which correspond to hydrostatic pressures several times greater than atmospheric. There
is a long and colorful history of the strife incurred in search-
ing for the exact state of water in these structures, and as
the sap rising problem is so pertinent to discussion of osmotic
pressure and imbibition we shall recall a few of the highlights.

In his classical essays "Vegetable Staticks" of 1727 (32), which
was five years later followed by a "Haemastaticks", The Reverend
Stephen Hales was preoccupied with measuring the forces of circulation in plants and animals. He states:

> "Tho vegetables (which are inanimate) have not an engine, which, by its
> alternate dilatations and contractions, does in animals forcibly drive
> the blood through the arteries and veins; yet has nature wonderfully
> contrived other means, most powerfully to raise and keep in motion the
> sap, as will in some measure appear by the experiments in this and the
> following chapter."

He demonstrated his thesis in a variety of plants. Some of the
more lucid experiments dealt with grapevines *(Vitis vinifera)*.
In the early spring before the leaves are out twigs bleed when
broken, and he carefully connected the cut stems with mercury
manometers and registered positive sap pressures. His highest
reading on a stump severed near the ground amounted to 38 inches of mercury, or about 1.3 atm.

> "Which force is near five times greater than the force of the blood in
> the great crural artery of a Horse; seven times greater than the force
> of the blood in the like artery of a Dog; and eight times greater than
> the blood's force in the same artery of a fallow Doe: Which different
> forces I found by tying those several animals down alive upon their
> backs" and then laying open the great left crural artery, where it first

enters the thigh, I fixed to it (by means of two brass pipes, which run
one into the other) a glass tube of above ten feet long, and 1/8th of an
inch diameter in bore: In which tube the blood of one Horse rose eight
feet, three inches, and the blood of another Horse eight feet nine inches.
The blood of a little Dog six feet and half high: In a large Spaniel seven feet high. The blood of the fallow Doe mounted five feet seven inches."

Stephen Hales
Rector of Farringdon, Hampshire
Minister of Teddington, Middlesex

Mercifully - let us return to "Scientia amabilis", as botany was
affectionately called by the temporary Linnéan school. Hales,
in a great variety of experiments on twigs and roots of growing
plants with the leaves fully out, found that water was sucked
in with a force as high as 10 to 12 inches of mercury, or nearly
half an atmosphere:

"From these experiments I say it seems evident that the capillary sap vessels, out of the bleeding season, have little power to protrude sap in
any plenty beyond their orifices; but as any sap is evaporated off, they
can by their strong attraction (assisted by the genial warmth of the sun)
supply the great quantities of sap drawn off by perspiration."

In the early 1890's Eduard Strasburger addressed himself square-
ly to the central question: how can a tree transport sap several
times higher than corresponding to the atmospheric pressure of
10 m ? He demonstrated in a large series of tall trees that sap
is *pulled* to the leaves, not *pushed*, and no living pumps are
involved (95, 96). The trees, some 20 in all and of varying

species, ranged in height from 11 to 22 m, i.e. all were in the range exceeding barometric pressure. They were tied by ropes to neighboring trees and lifted by block and tackle, as they were sawed off at the base with water flushing the saw. Dangling, with the base in a bucket of water, the doomed tree was given half an hour's grace, before poison, e.g. copper sulphate or picric acid, was stirred into the bucket. The brew traveled up the stem to the very top in a few days. Sometimes over 100 l were consumed in one week.

The botanical laboratory was housed in the stately Poppelsdorfer Castle, festooned with gorgeous wistaria vines, up to 20 m tall. Several of these were chopped off at the base and presented with selected poisons. The stem of one of them was forced into a loop, which was dunked in a bucket of $90^{\circ}C$ water for half an hour; but neither poisoning, nor cooking kept the sap from rising!

By careful execution of these heroic experiments Strasburger closed the gate for all time on a vitalistic theory of sap rising, and triggered the efforts to find a purely physical explanation: Within a year such was already forthcoming.

Eduard Strasburger
University of Bonn

Henry Dixon
Trinity College
University of Dublin

In 1894, H. Dixon and J. Joly (22), and independently E. Askenasy (6) shortly thereafter, launched the hypothesis that sap rising depends upon a cohesive system where fluid water under tension is stretched from the soil, through root and stem, to the leaves, with evaporation from the leaves supplying the driving force. This is the celebrated cohesion theory for the ascent of sap, aptly called also the wick theory.

In his major opus of 1914, "Transpiration and the ascent of sap in plants", Dixon (21) found that a technique similar to Berthelot's (p. 12) yielded a tension as high as -150 atm in air-saturated water, and up to -203 atm in sap from ivy (*Ilex*). He also observed that little specks of wood added to the water did not affect the tension, which evidently was limited by adhesion to the glass rather than to the wood. But, although it was thus established that both water and sap were able to sustain ample tensions *in vitro*, there was still no technique to measure exactly the state of the sap in trees and vines.

Dixon's main concern was whether there was adequate correspondence between the osmotic pressure in the leaf cells and the height, with the added considerations of flow resistance and turgor. He first tried to estimate the osmotic value with DeVries' plasmolytic technique, which did not work out:

> "There are several reasons why the application of this method is not suitable to leaf cells. In the first place, it is necessary to cut sections of the leaf in order to apply the solutions and to allow microscopic observation. The injury involved in sectioning acts as a violent stimulus to the tissues, which may in itself evoke a change in the concentration of the vacuoles or contraction of the protoplasm. Secondly, accurate determination of the plasmolyzing concentration is very difficult, as the contraction of the protoplasmic membrane must be considerable before it can be observed microscopically."

He then turned to "Osmotic pressure balanced by gas-pressure", and for this purpose he constructed a cylindrical glass chamber 1 cm thick, 50 cm tall and 10 cm wide (Fig. 17). Lid and bottom were metal castings pressed air-tight against the glass by gaskets and three slender bolts. Air pressure was supplied from a pump, but "as the pump I had at my disposal was unable to compress air above a pressure of about 10 atm, I discarded it in favor of a bottle containing liquid carbon dioxide..." and then with a dig at a measly budget he continues: "This has the additional advantage that careful observations are possible while raising the pressure, which can not be done while using a pump unless an assistant is employed."

He outlines an experiment as follows:

> "A branch bearing a number of leaves is enclosed in a strong glass cylinder, capable of resisting high gas-pressure (e.g. 50 to 100 atm), and the pressure is raised in this vessel by means of an air-compression pump, or by attaching it directly to a cylinder containing liquid carbon dioxide. The lower portion of the branch projects from the cylinder and dips into a glass vessel containing a weighed quantity of water. These arrangements are shown in Fig. 24. It is evident that when the gas pressure in the glass vessel surrounding the branch is raised and maintained above the osmotic pressure of the cells of the leaf, water will be forced from these cells back into the conduits of the branch and into the vessel beneath. This will become apparent in two ways: first by *flagging of the leaf*, inasmuch

as the rigidity of the leaf is due to the internal pressure of these cells, so that when this pressure is overcome by the external gas-pressure the leaf will flag; secondly, by increase in weight by the vessel beneath containing the water into which the branch dips."

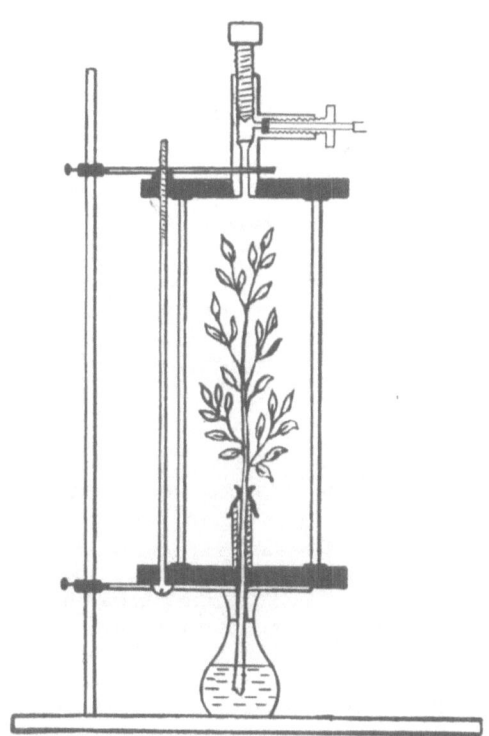

Fig. 17. Dixon's pressure chamber. Glass cylinder 10 x 50 cm, 1 cm thick (his fig. 24). He experienced two explosions with it and gave it up (21)

In these determinations the primary readout was incipient wilting or "flagging" and hence required a transparent chamber. He experienced a number of difficulties with the toxicity of high $CO_2$ pressures; his highest figures were obtained from Linden (*Tilia americana*), 38 atm.[4]

During this interlude with the pressure chamber, Dixon evidently failed to realize that a properly applied, and properly observed, balancing pressure would yield the instant sap pressure, i.e. the osmotic pressure in the leaf cells minus the turgor pressure. His flagging pressure, it is true, gives a rough indication of the osmotic pressure at wilting, obtainable in some

---

[4]*Explosions*. Of these experiments he states:

"...by means of this method useful results were obtained, but danger attended the determinations. Despite the strength of the glass cylinders used, two explosions occured, fortunately attended by delay in the work only, so that after a comparatively small number of observations, a more suitable method was looked for."

plants, not in others and the weighing technique, if at all possible, is clearly not the way. Nevertheless, considering his remarkable general insight, he had a near miss to measuring the
sap tension in tall trees, a goal reached 50 years later (p. 31).

His final technique, and the most successful in determining the
osmotic pressure, utilized freezing point determinations in
juices pressed from leaves which had been macerated by freezing
in liquid air. Calculated osmotic pressures were generally much
in excess of those demanded by transpiration and seemingly correlated more closely with the state of photosynthesis. He concludes:

> "The transpiring cells of the mesophyll normally remain turgid during trans-
> piration; accordingly we would expect if our line of reasoning is correct,
> that in high trees the osmotic pressure keeping them distended must corres-
> pond in magnitude to the tensions necessary to raise the sap. This surmise
> has been confirmed by determinations of the osmotic pressures of the sap
> of various leaves. These pressures have always been found adequate to re-
> sist the transpiration tension; but in many cases other factors enter in,
> and the pressures developed are much in excess of those demanded by trans-
> piration."

Dixon's ideas (1914, 21) on the mechanisms of the osmotic relations in the leaves were remarkably accurate

> "During transpiration the cells of the leaves are normally in a turgid
> condition. This distension is caused by the osmotic pressure of the dis-
> solved substances acting upon the protoplasmic membranes of the cell and
> pressing them against the cell walls"..."We may then regard secretion or
> evaporation as the force which actually exerts the tension on the sap,
> and this tension is transmitted through the leaf cells to the sap in the
> conducting tracts."..."*The simultaneous presence of pressure and tension*
> (our italics) within these cells, at first sight, appears paradoxical;
> but a moment's consideration will show that it is quite possible for the
> solvent, water, to be in a state of tension, i.e., at a negative pressure,
> while the dissolved substances may be at a positive pressure and be active
> as a distending force on the cell."

The latter, is what we call "dispersal pressure". It is always
present in any solution and manifests itself in osmotic as well
as turgor pressures under appropriate conditions.


Sap Tension by Rate of Filtration

In this creative period Otto Renner, at the University of Bonn,
came the closest to a direct, experimental insight into the sap
tension problem, using an ingenious rate-of-filtration approach
(1911, 73). He connected a leafy twig with a capillary and by
compressing the stem with a screw clamp he increased the flow
resistance and forced the leaves to exert a maximal transpira-
tional pull (Fig. 18). When the leafy top was severed and a
vacuum line was attached to the stump, the rate of flow dropped
to about one-tenth. The conclusion from this simple and direct
experiment seems inevitable, namely that the leaves must have
transmitted a pull of 10 atm to the sap. By slight modification

of the technique he was also able to show that negative pressure of a few atm is regularly found in twigs attached to the tree (74, 75).

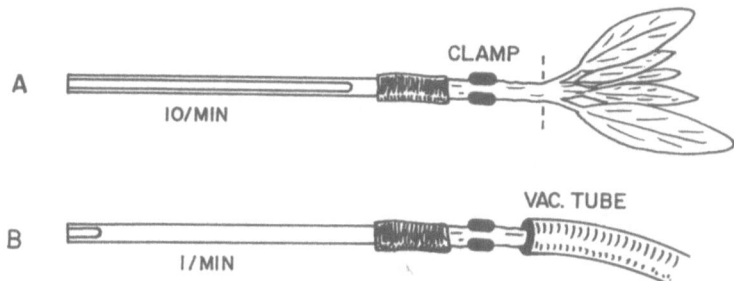

Fig. 18 A and B. Renner's rate experiment. (A) leafy twig with clamped stem, (B) vacuum tube drawing water through the same stem at a rate only one-tenth that of the leaves (87).

We adopted Renner's rate technique in various ways, but after a considerable amount of work we found ourselves unable to attain consistent results, in a quantitative sense. True enough, in mangroves, negative pressures were frequently indicated, but data later obtained by use of the pressure chamber showed the rate method to be generally far from the mark (84).[5]

## Gravitational Pressure Gradients in the Xylem

We have already demonstrated that in a variety of gels and matrices at equilibrium the only effect on the mobile water derives from forces developed by menisci at the free surface. There is, however, a growing notion that "matric potential", in one form

---

[5] *Rustlings in the forest.* At a dry faculty luncheon at Harvard, Professor Irving Bailey recalled to the writer: "Years ago when I could not believe the cohesion story I said to myself: 'If the story is right there must be some way to prevent air from spreading when a twig breaks off, there must be some relation between the pressure necessary to force air through a twig and the height of the twig above the ground.'" So in 1916 he studied twigs taken from a height of 25 m and found that air leaked through at 3 atm which was only half of the estimated pull during transpiration (8). He then told the following story:

"Years later Dr. MacDougal at the desert laboratory at Tucson invited me to see his famous dendrograph in action. Clamped to a tree it wrote a diurnal curve of the diameter of the trunk, shrinking by day, swelling by night, beautifully in line with the cohesion theory. After watching this for several days I suggested: "Now Mac, why don't you stick your gadget onto a telephone pole?" Mac did, and lo and behold, next day the same magnificent curve! This casts no aspersion on MacDougal's amply creditable dendrograph (47); for who among us has not been bugged by a "telephone pole" experiment!"

or another may interfere with, or even abolish the hydrostatic gravitational gradient in xylem capillaries (13, 64, 93). We shall therefore emphasize the experimental evidence for negative pressure and normal gravitational gradients in these structures.

Excellent data were obtained on grapevines (88). In the spring before leaves are out evaporation is minimal and the vines bleed when broken. This indicates positive sap pressure, and can be easily measured (Fig. 19). A special capillary manometer, closed in the end, is clamped airtight on the stem. It micropunctures the xylem, the sap compresses the known air column in the capillary and measures the pressure. As the gadget is moved from one level to the next it is mandatory that each puncture be instantly sealed to prevent general pressure drop from sap leakage. For a total height of up to 12 m the hydrostatic pressure *decreased* (Fig. 19) by very closely 0.1 atm per meter. No lift is derived from the walls. In another experiment the lower end of a 10-m section of a vertical grapevine was connected to a hose hanging next to it in a U formation. The system was flooded with water. As the top of the hose was raised or lowered the water receded or overflowed from the upper cut end of the vine, precisely as if the stem were a water-filled pipe. There is nothing in the xylem of the living stem which at equilibrium can violate the principle of Pascal, and thereby the Second Law.

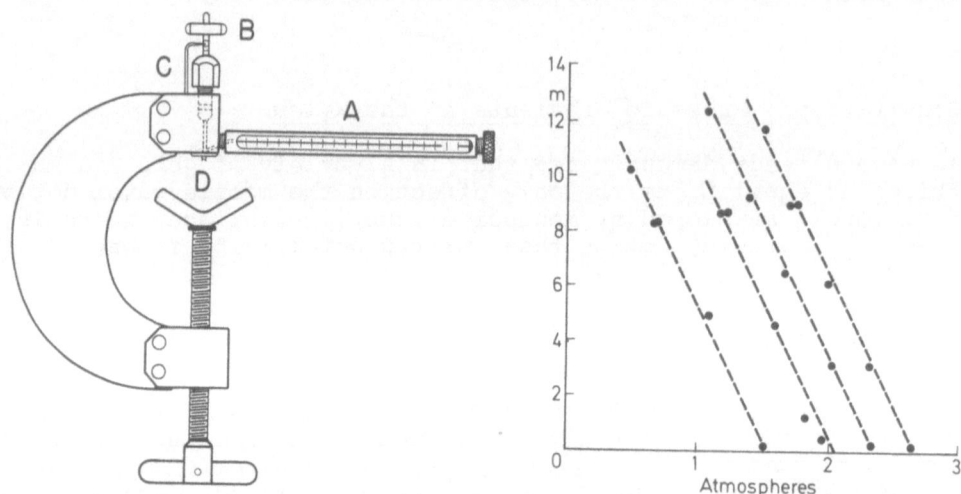

Fig. 19. Direct measurements of positive sap pressures in grapevines, *Vitis labrusca*, before the leaves are out. Left: instrument. *A* capillary manometer, *B* head of punch, *C* depth indicator. *D* contact button with air tight punch protruding. Screw clamp holds vine air tight against button. Right: hydrostatic gradients in four vines. Slopes are 10 m elevation per atm sap pressure (88).

Related to these experiments are others performed on the long and narrow climbing stems of rattan palms (87). The slender stems grow from a basal leaf rosette and reach up into the canopy of

forest trees some 30-50 m high. Evidently the stem grows faster
than the host tree, for loops 10-20 m long are commonly found
lying on the ground. A tall stem about 2 cm thick was looped
at the base, cut under water and connected with a burette, so
the drinking rate of the crown could be measured. The burette
was then filled to the brim with water and stoppered, but the
drinking kept on at an undiminished rate against near vacuum,
for no air entered from the xylem.[6] Higher up the water was
therefore under tension, and in the high crowns it must attain
at least some -10 atmospheres during transpiration.

A few millimeters of the stems of a dozen thin twigs of mangroves and other plants were frozen briefly *in situ* between
little pieces of $CO_2$ ice (79). They all wilted and died in a
matter of hours or sometimes a few days. Removal of a ring of
phloem by ring barking and wrapping it with tape showed no such
effect. It was concluded that cavitation within the xylem caused
by freezing became irreversible and left a permanent tensile
break in the sap transport. It is remarkable though that certain arctic evergreen species can indeed be frozen *in situ* with
no sign of increased resistance to flow subsequent to thawing
(34).

We shall in the next example deviate from the chronology and
anticipate the use of the pressure chamber described in the
next section.

In several tall redwoods *(Sequoia sempervirens)* and Douglas
firs *(Pseudotsuga douglasi)* a high and a low twig some 50 m
apart were shot down and immediately protected by vapor tight
bags. Within 10 min the sap pressure was determined in both.
The elevation of each twig was determined by triangulation.
The difference in their sap pressures divided by their difference in heights, was equal to an equilibrium hydrostatic pressure gradient within ±10% (84).

There is no experimental evidence for any mechanism in the xylem
which at equilibrium can violate the principle of Pascal - which
would in fact, automatically be a violation of the Second Law
of Thermodynamics (cf. p. 55-57, 71, 72).


## At Last: Measuring Negative Sap Pressure

In our search for a way of measuring sap pressure, mangroves
offered a choice material because they grow in a substrate which
may be well-defined. It is true that many reach their most lux-
uriant growth in estuaries, but some do well without access to
fresh water, and in these the roots are bathed in a solution
which is essentially sea water. Walter and Steiner (1936, 108)
had already suggested that the transpiration stream of mangroves
must be essentially fresh water, for otherwise there would be an
impossible accumulation of salts in the leaves. But, on the other

---

[6] The same effect had earlier been obtained on grape vines (88)

hand, in order to pull fresh water from the sea one would anticipate an osmotic pressure in the leaves at least as high as in
the sea, or rather considerably higher in order to preserve their
turgor. That mangroves and other halophytes have such a high concentration of salts in their cells was well known (86, 91).

The composition and pressure of xylem sap were unknown. Sap was
therefore collected by applying suction to one end of a stem section while clipping the other end down piece by piece (Fig. 20,
(9)). In certain mangroves like *Rhizophora* and *Laguncularia* it
turned out to be very nearly salt free, with a melting point of
$0.1^\circ$C or less (83, 86). This gave an osmotic potential between
sap and sea water of some 25 atm, and even higher gradient between sap and parenchyma cells. In other words the mangroves
should have a sap tension of at least -25 atm if the cohesion
theory were valid.

It would also follow that the separation of fresh water from sea
water could be considered a simple physical ultrafiltration i.e.
an analog to reverse osmosis, which is used by desalination in-
dustries. There would be one big difference, however: in the in-
dustrial process the water is *pushed* through a membrane, in
plants it would have to be *pulled* through the root membrane
(cf. Fig. 13). If therefore, an intact mangrove seedling were
placed in sea water inside a pressure chamber with the cut stem
sticking out (Fig. 21), it should be possible to apply high

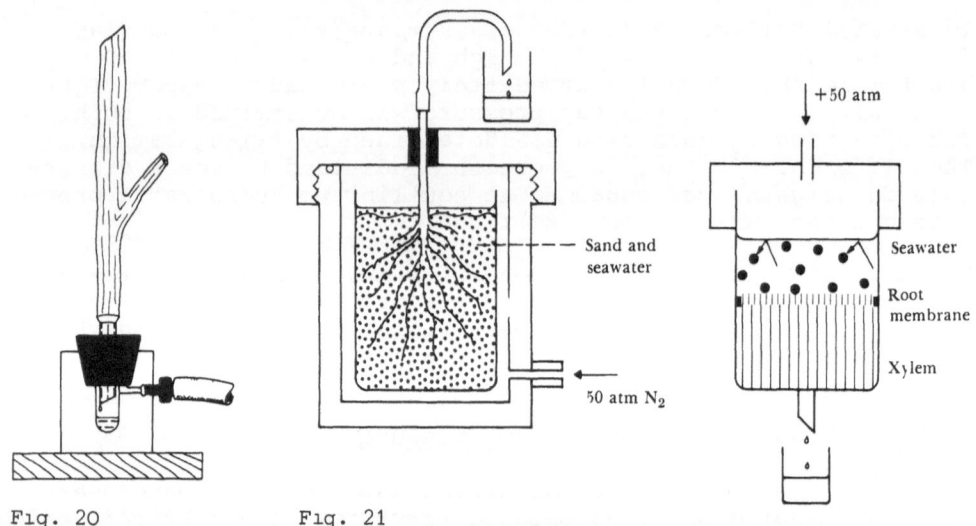

Fig. 20          Fig. 21

Fig. 20. Collecting xylem sap by vacuum as stem is clipped down, one small
piece at a time, after method of Bennet, Anderssen and Milad, 1927 (9).
This procedure destroys the valves so the sap can descent step by step

Fig. 21. *Left:* Decapitated mangrove seedling placed in a pressure chamber
produces a flow of fresh water when subjected to 50 atm pressure. *Right:*
The reverse osmosis displayed by the seedling is schematically presented
in an inverse position. An intact plant performs the feat by a 50 atm trans-
pirational *pull*, not by a *push* as in the Figure (82)!

enough pressure to produce an ultra filtration (reverse osmosis) from the roots.

Such experiments were tried on an expedition to Baja, California, February-March, 1964, and true enough, at some 50 atm gas pressure fresh water kept bleeding from the cut stem of one of several transplanted seedlings, while the rest leaked seawater (84). As it turned out, this technique had already been used in other plants by Mees and Weatherley (51) and others. This single observation was later verified and amplified on several Australian mangroves (79).

However, another event on our Baja expedition was to gain the limelight. The stage was replete with discussions and ideas on reverse osmosis; instrumentation was ready and waiting, and a breakthrough on the old sap-rising problem was imminent. Indeed, hardly without losing a day, colleagues Hammel and Hemmingsen got the idea of applying the chamber technique to mangrove *twigs*. Having mounted them in a small (100 cm$^3$ and safe!) pressure chamber, it was found that sap emerged at a very definite gas pressure which varied little amongst different twigs of the same plant.[7] It was soon realized that this must be *a direct measurement of the expanding force of the matrix*, and hence of *the negative hydrostatic pressure in the sap!*

The technique was immediately exploited and further developed, and demonstrated that all vascular plants tested had negative sap pressure, most markedly so: the mangroves and desert plants (Fig. 22, (77, 80, 84, 85, 100, 101)). We have already pointed out (p. 26, 27) that a pressure chamber was tried and discarded in 1914 by Dixon, as a tool for estimating osmotic parenchyma pressures.

One version of such a pressure chamber is shown in Fig. 23. With it one may obtain all, or part of, an inverse pressure-volume curve, or simply a *pressure-volume curve*. A leafy twig is enclosed within the chamber leaving the cut end protruding through an airtight seal into the open air (Fig. 23A). Pressurized nitrogen gas is admitted stepwise to the chamber until sap appears, and remains stationary in the cut end. This is the initial balancing pressure, and represents the negative hydrostatic sap pressure.

---

[7]Explosion. A few days later we, like Dixon, had a serious scare. Three of us sat clustered around a mangrove twig kept in a large, 4 l chamber made from an old discarded steel cylinder of dimensions similar to Dixon's. At 900 PSI (61 atm) the writer observed the cut end through a short focus hand lens but the sap had not yet surfaced. As he leaned back and stepped up the pressure, the chamber exploded, blowing out the bottom. Others saw the deadly missile shoot 10 m into the air, and incredibly, it fell exactly back in its place without hitting anybody. Bare legs were bleeding from flying sand and the eye glasses of the only one who happened to be looking in that direction were blasted opaque. Old material was not used for future chambers, only hydraulically tested new ones.

34

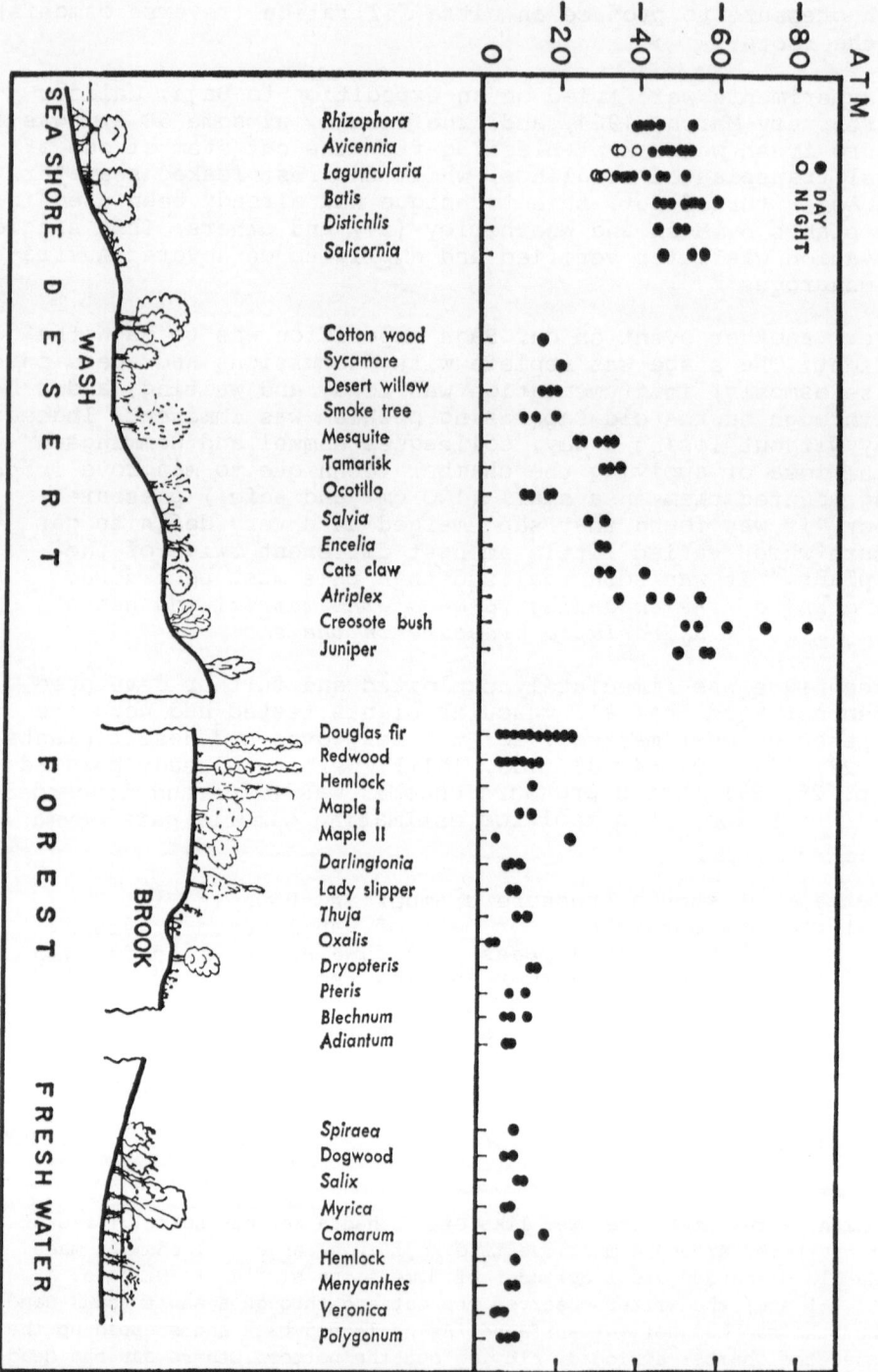

Fig. 22. Sap pressure in a variety of vascular plants. Most measurements were taken during daytime in sunlight. Night values are likely to be less negative by several atm (84)

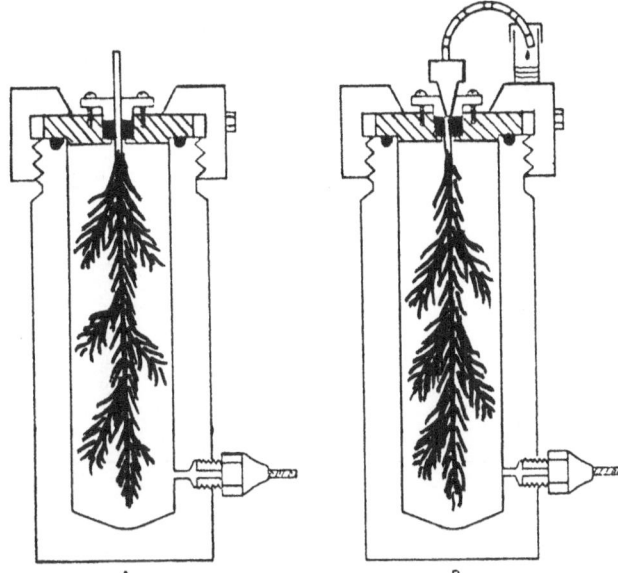

Fig. 23 A and B. Pressure
chamber for measurement of
sap tension in a twig.
*(A)* direct observation.
*(B)* step-by-step sap ex-
trusion and pressure mea-
surement to obtain a pressure-volume curve (80, 84)

A more sophisticated procedure gives a pressure-volume curve
(80, 84, 100, 101). A transparent collecting tube is fixed to
the cut end and measured quantities of sap are extruded
(Fig. 23B), each followed by observing, through the tube, the
new balancing pressure. When no more sap can be extruded, the
remaining extracellular water in xylem vessels and cell walls
is calculated from the dry weight of the twig.

Using a pressure vessel, even salt-secreting leaves deliver a
sap which is practically pure water, with a freezing point usually less than $0.01^{\circ}C$. Ideally therefore, at zero turgor there
should be a one-to-one correspondence between the osmotic pressure and the sap tension in the xylem vessels. If $J$ is the initial intracellular water, $V$ the eliminated water, $S$ the solutes
and $p$ the equilibrium pressure we may write the following equation:

$$\frac{S}{J - V} = Kp ,$$

or as $S$ and $K$ are constants this reduces to

$$K_1 \frac{1}{p} = J - V; \text{ where } K_1 = \frac{S}{K} .$$

If therefore we plot the inverse of the pressure $(\frac{1}{p})$ against
liquid removed $(V)$ we get a straight line whenever the concentration is proportional to the pressure. A schematic curve is
shown in Fig. 24. Indeed, the relation is linear until about
half of the initial intracellular water is extruded; usually
20-30% of the total water remains in a space that does not col-
lapse; including cell walls and xylem (84).

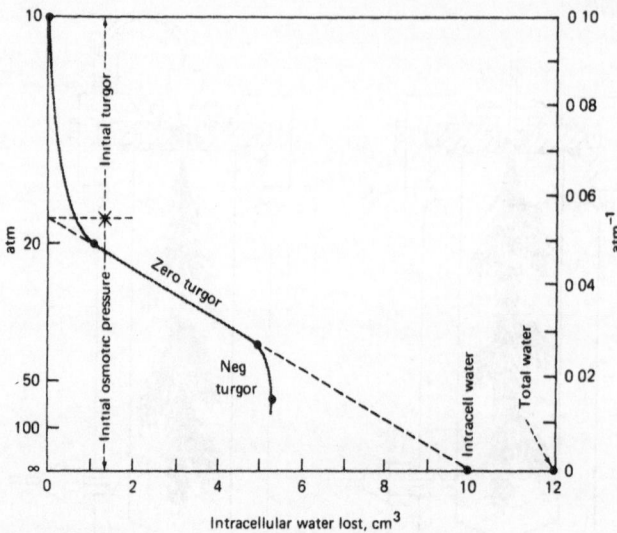

Fig. 24. Schematic presentation of a pressure-volume curve (80)

## Negative Solvent Pressure in Xylem and Cells

We now had separate techniques for comparing a) *the negative hydrostatic pressure of the sap in the xylem* with b) *the osmotic pressure within the leaf cells*, which should balance at zero turgor (wilting). Fig. 25 shows determinations made at zero turgor on a variety of halophytes and trees. The melting point of the sap was proportional to the equilibrium pressure which was measured at near $25^\circ$C, where $-1.86^\circ$C corresponds to an osmotic potential of $-24.5$ atm. The empirical data fell remarkably close to the theoretical line, showing that the leaves behave very nearly as perfect osmometers, complementing the fact that nothing but pure water is extruded.

This striking experimental correlation between *osmotic pressure* in the wilting leaf cells and the *negative hydrostatic pressure* in the xylem appears also in the same twig at different temperatures, as was first observed by Stanley Miller (personal communication) on the expedition of the R/V Alpha Helix to the Great Barrier Reef, 1966. He found that the balancing pressure of wilting mangroves *Lumnizera* and *Osbornia* indeed suggested proportionality to the absolute temperature, and H.T. Hammel (personal communication) later found the same close correlation in a conifer (Fig. 26). This of course means that we are measuring in the *salt free xylem sap* a direct expression of the thermal motion of *the solute molecules* in the leaf cells.

The mechanism is clear: The osmotic pressure is caused by the thermal dispersal pressure of the solute molecules within the cells. At zero turgor pressure the cell wall is slack and the dispersal pressure of the solute must therefore be absorbed by the boundary of the solvent. The *negative solvent pressure* thus created is transmitted through the semipermeable cell membrane into the pure water in the xylem, where it is measured *as such*

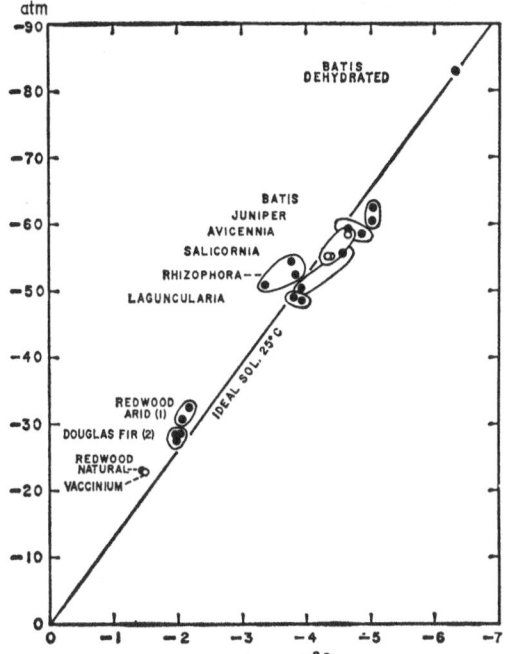

Fig. 25. Melting points of intracellular juices from frozen and
crushed leaves, in relation to
sap tension measured at zero
turgor. Melting points corrected
for extracellular water. Melting
point of $-1.86°$ corresponds to
24.5 atm solute pressure at $25°C$
(84)

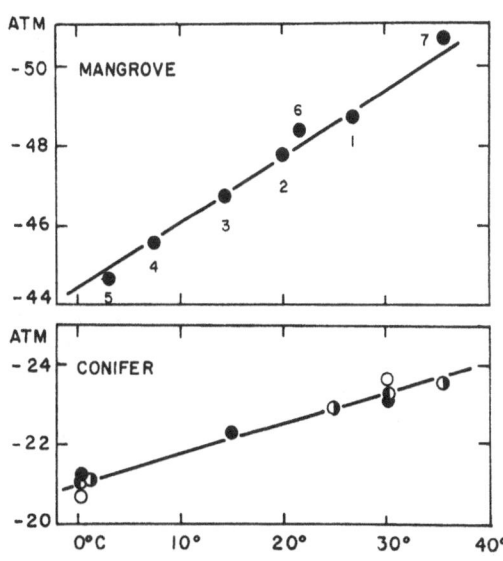

Fig. 26. Sap pressure of wilting vascular plants at a series of temperatures,
demonstrating proportionality with the Kelvin temperature *(diagonal)*, and
hence mirroring the kinetic pressure within the parenchyma cells. *Upper:*
One twig of mangrove *(Lumnizera littorea)*. Numbers are the order in which
the measurements were taken. S. Miller, 1966, unpublished. *Lower:* Four twigs
of conifer *(Cupressus* hybr.). H. Hammel, 1973, unpublished

by the pressure chamber. This is the essence of the negative
solvent pressure theory of osmosis.[8]

Let us scrutinize this idea in simple models. In Fig. 27A is a
limp leaf cell containing a molar solution. It is separated from
the xylem by a semipermeable membrane. The xylem consists of a
system of fine tubes which are prevented from collapsing by the
rigidity of the walls. The lower end is closed by a semipermeable membrane dipping into sea water of 24 atm osmotic pressure.
We found (Fig. 25) that if the solute concentration were 1 molar
(-1.86°C melting point) then the hydrostatic pressure at 20°C
of the salt free sap within the xylem would be -24 atm. Fig. 27B

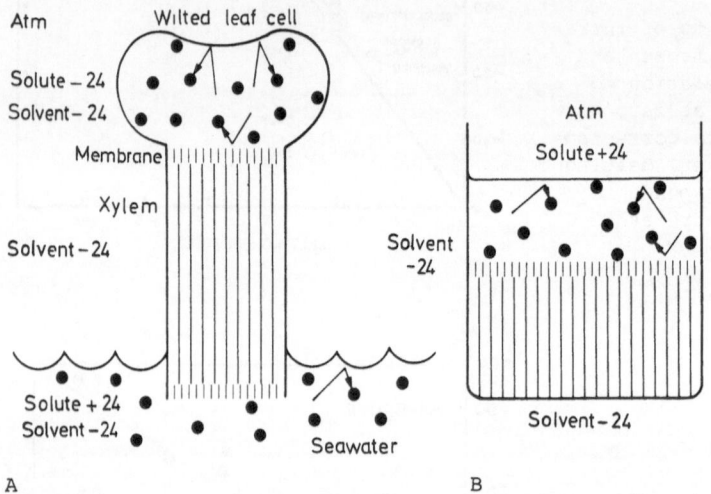

A                                          B

Fig. 27 A and B. (A) Simplified diagram of a low mangrove growing in seawater.
The system is at equilibrium with no turgor in the cell and there is no evaporation. The vapor pressure throughout is that of the tensile xylem sap,
which is the solvent. All parameters are measured directly except the hydrostatic pressure of the solvent in the cell and the seawater. (B) An analog
system. The solution has a free surface and rests on a semipermeable membrane.
The solvent below is prevented from cavitating by being held in a capillary
or matrix structure. Arrows signify the solute pressure of +24 atm generat-
ing -24 atm in the solvent through coupling at the free or unrestrained
surface (82)

---

[8]Dixon essentially proposed this view when in 1914 he wrote about turgor
pressure (21): "The osmotic pressure is the pressure which the dissolved substance exerts against the membrane of the cell while the tension is in the
solvent and is transmitted unaltered across the space in which the pressure
of the solutes is also exerted. In this respect the osmotic pressure acts in
just the same way as a number of internal supports keeping the cell turgid
and preventing it from collapsing under the tension of the solvent which
drags the water across the cell."

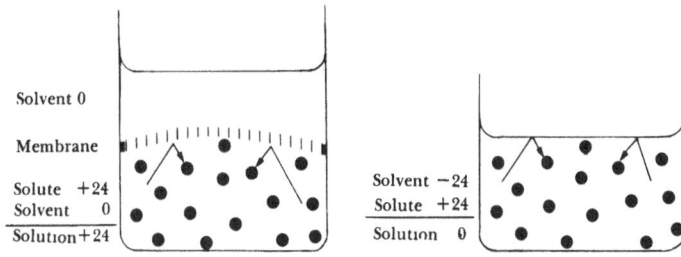

Fig. 28 A and B. (A) The thermal dispersal pressure of solute molecules bucking a semipermeable membrane which is fastened at the edge. Water goes freely through the membrane and is ambient throughout, and the pressure on the membrane is caused by the solute molecules alone. (B) The same concentration of solute molecules now bucking the free surface and lending negative pressure to the solvent. Solutions are 1 mol at 20°C (82)

shows an analog system *in vitro*, where the solute molecules, straining against the surface put the solvent under a negative pressure which is transmitted through the membrane to the matrix below.

Leaving the plant model, let us turn to a molal solution below a semipermeable membrane with water above (Fig. 28A). This membrane is found to be strained by +24 atm which is the osmotic pressure at 20°C; why so? It is due to change of the momentum as solute molecules are reflected from the semipermeable membrane (40, 55), i.e. van't Hoff's bombardment pressure.

However, if we put the solute molecules behind another barrier, namely, the free water surface (Fig. 28B), they are again reflected but now they exert their force of +24 atm on the free surface, and this of course lowers the hydrostatic pressure of the water to -24 atm; or as Professor Hugo Theorell in Stockholm expressed it: "Of course! Solute molecules are blind and don't know what they hit!"

In summary: the kinetic force of the dispersing solute molecules is absorbed by the cohesiveness of the water, and at equilibrium the force coupling is the free surface: from here the negative pressure is transmitted throughout the solvent. The solute pressure is +24 atm, the solvent pressure is -24 atm and the solution pressure ambient. The parallel to tensile water in a compressed matrix is perfect!

What other criteria do we have for this lowering of the solvent pressure? Indeed, that the vapor pressure goes down by an amount exactly corresponding to a mechanical change in pressure for which Poynting wrote the thermodynamic expression in 1881 (65). So here then, we have full insight into the lowering of the vapor pressure over a *solution*, a subject which thermodynamicists left by the wayside, under the banner of an activity effect of the solute molecules.

A physical explanation for the lowered vapor pressure which is sometimes offered, is that the solute molecules interfere with the exit of the vapor by blocking off part of the solvent surface. However, at equilibrium exit- and entrance rate of the vapor must be the same, and distillation would ensue if partial blocking interfered with the equilibrium vapor pressure. Floating particles or oil films retard the vapor exchange but do not change the equilibrium.

## Pressure Profiles through an Osmotic Model System

The ideas embodied in Figs. 27 and 28 dealing with equilibria may be presented schematically in Figs. 29 A and B. In A, the solute molecules are confined between the stiff shell and the rigid membrane and have no effect on the water, but only on the wall and the membrane. The pressure in the solution is hence

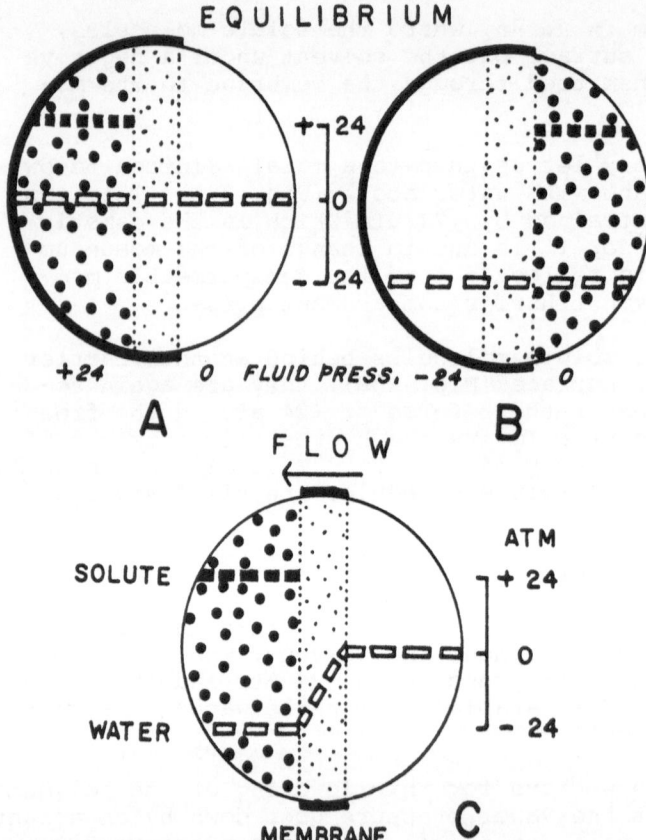

Fig. 29 A - C. Osmotic pressure profiles across a semipermeable membrane in small weightless systems with spherical free surfaces. Membranes and shells are stiff. Left halves of upper systems are closed and there is no flow. The lower system has two free surfaces and flow. According to the tension theory, the membrane can sustain no gradients of the solvent at equilibrium and the pressure profiles run accordingly. (A) and (B) are equilibrium states, (C) is not (81)

+24, water 0, solute +24 atm. In B the solute molecules strike
the free surface and put the solvent in the entire system under
-24 atm, the solute is +24 and the solution hence ambient. The
vapor pressure in B is lowered to correspond to -24 atm water
pressure according to Poynting's relation. In the open system C,
water flows through the membrane into the tensile water of the
solution, either by hydraulic or diffusive flux, depending upon
the pore size of the membrane.

Profiles A and B are consistent with the idea (which is obvious
for matrices), that pure water, although *interrupted* here and
there by a solute molecule, forms a *continuous* web which stret-
ches from the solution surface, through the porous membrane, to
the solvent surface. In this continuum of pure water the hydrostatic pressure propagates undiminished in all directions, obey-
ing the principle of Pascal. The empirical fact that the surface
impact of a solute molecule does not simply *substitute* for that
of a water molecule, but increases the tension in the water, is
the basis for this simple model, and as we shall see, consistent
with the Kelvin-Poynting gravitational column, i.e. solute and
solvent act as independent kinetic units.

## C. Solutions and the Role of the Boundary

### Solvent Tension and the Curved Surface

The statement is commonly accepted that the chemical potential
of water in a beaker is lowered by solutes and that the *reason
for this* in many applications does not matter. To the more physically minded, however, this is not a satisfactory state of af-
fairs and in retrospect it is interesting to note that at least
half a dozen investigators independently arrived at solvent tension as the natural explanation for osmotic phenomena, long before we got the clue from our mangrove studies.

In 1871, Sir William Thomson, Lord Kelvin (98), deduced that
the relation between the rise of water in a capillary and the
lowering of the vapor pressure is a simple inverse function of
the diameter (p. 43). According to his basic and elegant reasoning pertaining to an isothermal, evacuated enclosure (Fig. 30),

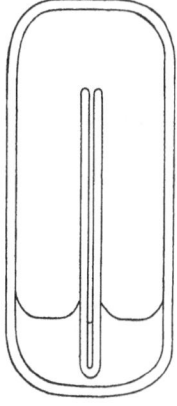

Fig. 30. Lord Kelvin's capillary column, 1871. In due
time the capillary would fill up by distillation so the
meniscus would reach exactly the height it would have
if the capillary were open below. Chamber evacuated and
isothermal (98)

Sir William Thomson
Lord Kelvin
University of Cambridge

John Henry Poynting
University of Cambridge

the vapor pressure of the curved meniscus at equilibrium must
match the gravitational attenuation of the vapor emanating from
a flat free surface

"if, as illustrated by the annexed diagram, a capillary tube with a small
quantity of liquid occupying it from its bottom up to a certain level be
placed in the neighborhood of a quantity of the same liquid with a wide
free surface, vapor will gradually become condensed into the liquid in
the capillary tube until the level of the liquid in it is the same as it
would be were the lower end of the tube in hydrostatic communication with
the large mass of liquid."

He ascribed the lowered vapor pressure in the capillary to the
curvature of the meniscus.

John Henry Poynting (65), working in Kelvin's laboratory, in 1881
humbly suggested that the cause of the change in the vapor pressure was the change in hydrostatic pressure of the water, independently of curvature *per se*:

"Sir William Thomson has shown...that if a liquid rises in a capillary
tube so that its surface is concave upwards, and (we may add) the pressure of the liquid is less than at the plane surface, then the equilibrium vapor-tension is less than at the plane surface. If the liquid
falls in the tube, so that the surface is convex and the pressure greater than at the plane surface, then the equilibrium vapor-tension is
greater. It has been supposed that this difference of vapor-tension is

due to the curvature of the surface...I think that we must look for the explanation elsewhere than in the curvature of the surface; and I shall endeavor to show that we may account for the effect by the difference of pressures of the liquid at the curved and plane surfaces. *The curvature of the surface is then, as it were, an accidental accompaniment of the difference of pressure, and not the cause of the variation in vapor-tension. We might therefore expect to find the variation taking place also at flat surfaces if the pressure be altered, and with solid as well as with liquid bodies.* (our italics) We can not directly investigate the vapor-tension of flat surfaces under pressure; but I shall assume that we may here take, instead, the rate at which exchange takes place when the solid and liquid are in contact with each other."

Poynting continues:

"Sir William Thomson's formula is

$$p = \varpi - \frac{2T\sigma}{r(\rho-\sigma)} \quad , \quad \ldots \ldots \ldots \ldots \quad (5)$$

where

$p$ is the vapour-tension in contact with the concave surface,
$\varpi$ is the vapour-tension in contact with the plane surface,
$T$ is the surface-tension of the liquid,
$\rho$ and $\sigma$ the densities of the liquid and its vapour respectively
$r$ the radius of curvature of the curved surface.

If $P$ be the difference between the hydrostatic pressures just beneath the curved surface and just beneath the plane surface, equation (5) may easily be put in the form

$$p = \varpi - P\frac{\sigma}{\rho} \quad , \quad \ldots \ldots \ldots \ldots \quad (6)$$

or a pressure $P$ in the liquid increases the vapour-tension by an amount $P\frac{\sigma}{\rho}$."

In modern terms the Poynting equation is often written as $dP_1\bar{V}_1 = dP_g\bar{V}_g$; where $P$ is pressure, $_1$ is liquid phase, $g$ is gas phase and $\bar{V}$ is the molar volume, i.e. liters/mole (or density as $cm^3$/g molecule rather than $cm^3$/g).

As a numerical example we may take: when water at $20^{\circ}$ is subjected to a pressure change of +24 atm its vapor pressure of 17.5 mm Hg changes to a good approximation by 0.31 mm, in the same direction according to

$$24 \cdot 760 \cdot 18 = \Delta P_g \cdot 24000\,\frac{760}{17.5}; \quad \Delta P_g = 0.31 \text{ mm} .$$

Conversely, when vapor pressure over a free surface is lowered by 0.31 mm Hg the hydrostatic pressure of the water is -24 atm.

Discussing an imaginary vapor pervious duct submerged vertically from the surface, Poynting continues:

"In a quantity of liquid at a uniform temperature, the number of molecules interchanged across a surface will increase as we descend, owing to increase of pressure. If near the surface the number will be proportional to the vapour-tension at the surface, then at any depth the number will be proportional to the pressure in an atmosphere of vapour at that level which, at the level of the surface, has the pressure of the vapour in equilibrium; that is, the liquid will behave as a non-vapourizing solid through whose interspaces the vapour can move freely."

Let us paraphrase in more general terms the essence of Kelvin's and Poynting's deductions: at equilibrium isothermally enclosed matrices or solutions that are pervious only to water and/or its vapor, have a vapor pressure which matches, level by level, the Boltzmann distribution of vapor above or below the flat surface of the outside pure water.

## The Kelvin-Poynting Gravitational Column

In one form or another Lord Kelvin's column in Poynting's interpretation soon became recognized as a stalwart guardian of the Second Law. In 1888, Gouy and Chaperon (31) applied it to an osmotic column at equilibrium (Fig. 31). They state:

"One may suppose that there is no transfer of solvent across the membrane C, nor any transfer through distillation between the free surfaces $A_1$ and $A_2$..., i.e. osmotic equilibrium across the membrane coexists with distillation equilibrium between the free surfaces."

They conclude that the height $h$ of the osmotic column depends uniquely on the difference in vapor pressure at the free surfaces, not upon the shape of the osmometer, nor upon the location of the membrane, indeed, this conclusion would require no knowledge of the laws of osmosis nor their relation to concentration. P. Duhem (24) takes strong issue with this paper, pointing out that true equilibrium is not attainable. He furnishes lengthy equations for non-equilibrium states - no doubt nearer to reality, but missing the main point.

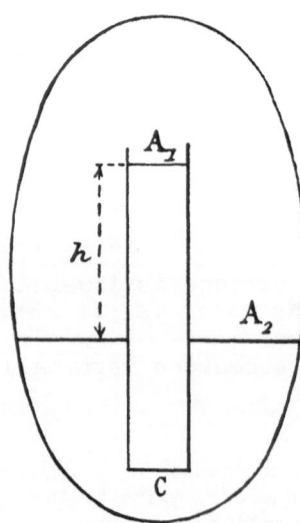

Fig. 31. The osmotic column by Gouy and Chaperon, 1888. C, semipermeable membrane. $A_1$ solution surface; $A_2$ water surface; $h$, height of osmotic column. Chamber evacuated and isothermal (31)

Fig. 32. Arrhenius' osmotic column, 1889. $L$ is osmometer with
semipermeable membrane at bottom, and solution rising to $h$.
$F$ is water with free surface
at $G$. Chamber evacuated and
isothermal (4)

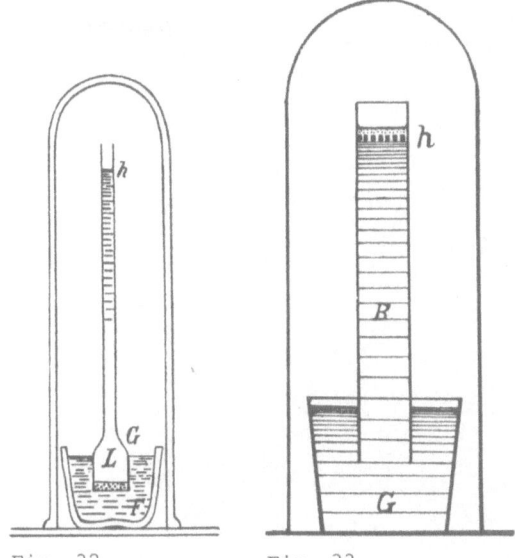

Fig. 33. Noyes' osmotic column,
1900. Semipermeable membrane at
top $h$, covered with an infinitely thin layer of solution. Ev-
erything below is pure water.
Chamber evacuated and isothermal (56)

Fig. 32                    Fig. 33

Arrhenius, 1889 (4), hails the simple elegance of the Kelvin-
Poynting column argument and considers a solution resting on a
semipermeable membrane at the bottom of a cylinder which is
dipped into a dish of water, the whole thing enclosed in air-
free isothermal conditions (Fig. 32). At equilibrium the vapor
pressure over the solution at $h$ must match that rising from the
pure water surface $G$; otherwise distillation would take place,
in violation of the Second Law. This would hold true for any
osmotic pressure, he thought, *provided* the densitiy of the solution were near 1.

Arthur A. Noyes, in 1900 (56), carried the column argument one
fundamental step further (Fig. 33). He placed the semipermeable
membrane on top of the column at $h$, rather than below, and by
covering it with an *"infinitely thin layer"* of solution he dis-
pelled the enigma that had worried Arrhenius, namely the density
of the solution. In this configuration[9], at least, it would seem
inescapable, to conclude, as Noyes did, *that the solvent is under negative pressure equal to $h$*, independent of the solute.
He deduced this also from a thermodynamic reversible cycle, and
one cannot but wonder how latter day thermodynamicists have managed to ignore the validity of his simple claims.


## Hulett and the Solvent Tension Theory

The major breakthrough of the osmotic mechanism was accomplished
by George Hulett in 1902 (41). With great detail and clarity he
develops the thesis of negative solvent pressure and indeed,
the essentials of the theory which we shall elaborate as the

---

[9]It is, in fact, immaterial where in the tube the membrane is located
(31, 81)

Arthur Noyes
Massachusetts Institute
of Technology

George Hulett
Princeton University

solvent tension mechanism of osmosis. As his paper remained vir-
tually unnoticed, we shall quote from it (in translation) rather
extensively, but shall first describe in a few words the circumstances which brought it about (97).

George Hulett graduated from Princeton in 1892, five years after
van't Hoff had announced his great discovery relating osmosis
to the gas laws, and went to Leipzig to work in Ostwald's laboratory. Wilhelm Ostwald, famous as a pioneer of physical chemistry, was a strong admirer and friend of the great theoretical
physicist Ernst Mach. It seems that neither of them at that time
believed in the reality of molecules, but rather considered them
useful abstractions (5, 61). In this superb milieu Hulett studied
for two years, earning his doctorate with a basic paper on liquid
crystals. Returning to his homeland he spent a few years at Ann
Arbor, Michigan, where he wrote a fifteen-page paper (41) on
"The relation between negative pressure and osmotic pressure".
It was accepted in the prestigious *Zeitschrift für physikalische
Chemie*, edited by Ostwald and van't Hoff, but to little avail,
Hulett's simple message all but faded away into oblivion.[10]
This was 1902, seven years before the atomic theory was convinc-
ingly propounded to the world by Jean Perrin. In 1909 Hulett was
appointed the first professor in physical chemistry in the United

---

[10]It was granted half a sentence in the last paragraph of Findlay's contem-
porary 77-page monograph "Osmotic Pressure" (27)

Jean Perrin
University of Paris

States at Princeton, and devoted himself, until his retirement
in 1935, mostly to bituminous chemistry, electro chemistry and
the theory of solubility.

Here is how Hulett, at the age of 35, sums up his solvent tension theory (41):

"When a solute is being dissolved in a solvent it diffuses under influence of the osmotic pressure in all directions to the boundaries of the
solvent. *There and only there is the solute stopped from further progress.*
When equilibrium is reached and the concentration is uniform there remains
an outward pressure by the solute on all points of the fluid boundaries.
This pressure tends to enlarge the fluid volume but it is checked by the
cohesiveness of the fluid. *One may therefore consider the osmotic pressure
as a negative pressure on the solvent,* (our italics) and a solution in
equilibrium with its water vapor represents a system of unequal pressure
on its two phases; for the pressure on the fluid phase is the sum of a
positive vapor pressure and a negative osmotic pressure. The sum is always less than the vapor pressure and is, as a rule, greatly negative.
*From these deliberations one may see why the vapor pressure of the sol-
vent decreases when a solute is added, namely because the osmotic pressure
renders the solvent pressure negative,* (our italics) and as we have showed
a mechanically induced negative pressure lowers the vapor pressure. The
quantitative relations between negative pressure and lowering of the vapor pressure are derived from a circle process which permits the calcula-
tion of one from the other, either way."

His isothermal cycle method, in fact, led to the already known
Poynting relation (p. 43).

In developing his solvent tension theory Hulett became intrigued
by the fashionable sap-rising issue. In fact, even before he
knew about the Dixon-Joly and Askenasy papers, he speculated
that negative pressure and cohesion must be the key to the problem. Askenasy (7) had already demonstrated cohesive lift by evaporation in a model consisting of a long-stemmed glass funnel
closed by a gypsum plate. The system was filled with water and
stood in a shallow dish of mercury. As evaporation proceeded
the mercury rose above atmospheric pressure before cavitating
the water. Unaware of this, Hulett built a similar model but
impregnated the gypsum disk with copper ferrocyanide. The mercury rose 111 cm. Thermostating the funnel and directing a rapid
air stream over the enclosed disk he observed that the rise of
mercury, i.e. the evaporation, slowed with increasing tension.
He considered this to be a crude, though first demonstration
of Poynting's relation. However, it appears from his data that
the rate of decrease in evaporation was at least 100 times greater than could be accounted for by the negative pressure.

Meanwhile, A. Noyes (56) had already claimed negative solvent
pressure which Hulett acknowledged, adding:

"We can go even further, as we have attempted to do, *namely to consider
the negative pressure imposed by the solute on the solvent to be the cause
of the lowering of the vapor pressure,* (our italics) for we have already
demonstrated that the vapor pressure is lowered by a mechanically induced
negative pressure."

Too bad that he was caught in a "telephone pole" experiment
(p. 29) on the last point! This, however, does not cast the
least shadow on the thermodynamic conclusion.

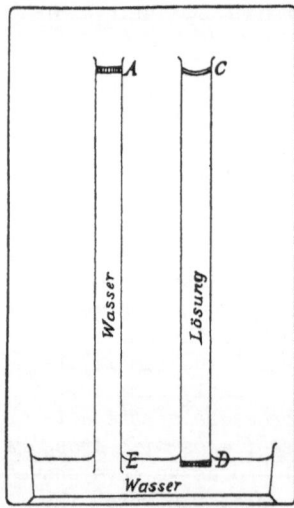

Fig. 34. Hulett's Figure 3 presenting two isothermal columns in a chamber containing only water vapor. *D* is a semipermeable membrane, *C* is
free surface of solution (should have been
flat). *A* is a porous membrane (41)

Using the gravitational column (Fig. 34) Hulett stated:

"Fig. 3 clarifies the relation between negative pressure and osmotic pressure. In tube CD is a solution and D is a semipermeable membrane. The solute presses outward on all solution boundaries and on the membrane. Consequently the volume of the solution ought to increase by this pressure but this is prevented by the cohesion of the solution and can only come about by entrance of solvent through the membrane D. The wall pressure of the tube is simply the hydrostatic pressure of the solution, and solvent enters at D *until the weight of the solution* (our italics) equals the osmotic pressure. It is clear from papers by Gouy and Chaperon and by Arrhenius that the vapor pressure at C is less than at D by the weight of the vapor column DC, i.e. when the system is at equilibrium; otherwise perpetual motion would be possible. Let us now take a similar tube, open below but covered above at A with a semipermeable membrane, freely pervious to the water vapor. The water wets the membrane and walls of the tube and is maintained by its cohesion. As both systems are kept under vacuum, the fluid in the left tube must be under negative pressure...At equilibrium the vapor pressure at A is less than at E, namely by the weight of the vapor column AE; i.e. the vapor pressure at the top of either column is the same when the height is the same. From this it follows that the negative pressure at A (which carries the water column) must be the same as the osmotic pressure which carries the solution, provided that one takes into account the difference in specific gravity of the columns."

It is clear from the following statement that Hulett intuitively subscribed to independent random movement of solute and solvent and hence to the validity of Pascal's principle for the solvent.

"It is evident from Fig. 3 (our Fig. 34. Authors' note) that the osmotic pressure must support a solution column of equal height as the water column provided the osmotic pressure and the negative water pressure are to produce the same vapor pressure lowering. Usually, however, there is a difference in densities between the two columns and the osmotic pressure at C is measured by the weight which is carried at D. *When the solute, as is commonly assumed, is independent of the solvent then the solute pressure on membrane D is greater than at C, namely by the weight of the solute.* (our italics). It follows that the weight carried by the osmotic pressure at C equals the solution weight minus the solute weight"

or in simpler words: the solvent at C is equally negative as at A! He continues:

"Osmotic pressure abides empirically by the laws of Boyle and Gay Lussac and one is therefore inclined to admit solute-solvent independence, but as Nernst points out this can not so far be decided."

Guldberg (1870, 30) had pointed out that ice and water in any solution at the freezing point have the same vapor pressure, and Hulett emphasizes that the freezing point of -0.187°C of a 0.1 molal solution matches that of pure water kept at -2.24 atm, for at that temperature the ice at 0 atm and the water at -2.24 atm have the same vapor pressure.

Hulett continues:

"It can also be shown that a given osmotic pressure results in the same lowering of the vapor pressure as an equal negative pressure, when one

considers the results of the lowering of the freezing point. It was shown
by Loomis, Raoult and others that a 1/10 norm. watery solution of an un-
dissociated solute had a freezing point of -0.187°C."

"Below 0°C the vapor pressure curve of the ice runs 0.044 mm lower per
degree than that for water; and the difference in vapor pressure between
the curves at -0.187° is 0.00824 mm. Consequently, a 1/10 norm. watery
solution with a (presumed) osmotic pressure of 2.24 atm. would have a
vapor pressure lowering of 0.00824 mm. We now ask, how great a vapor pressure lowering is produced by a mechanical negative pressure of 2.24 atm.?

From the above presented equation $-\Delta p = \dfrac{P\,\wp}{V}$ where $P$ equals -2.24 atm, $\wp$ the
volume of 18 g water, and $V$ the volume of 18 g water vapor at 0° at
4.569 mm, we calculate

$$-\Delta p = \frac{2.24 \cdot 760 \cdot 18}{22.4 \cdot 1000 \cdot 760/4.569} = 0.0082 \text{ mm,}$$

which accords with the above result." (Proof-reading mistakes corrected
by authors).

According to Hulett's argument the solute pressure induced a
negative solvent pressure which in turn lowered the vapor pressure as well as the melting point. Hulett had no valid experimental proof that the solvent in a solution was under tension;
but *he clearly recognized that the weight of the indirect evidence for solvent tension could quantitatively account for all
the colligative properties of the solution.*

Hulett carried his reasoning as far as experimental fixes would
permit and formalized it within the bounds of carefully applied
thermodynamics. A basic hypothesis was *solute-solvent independence*, which meant that within the solution at equilibrium there
could be no interference with respect to random movement, barring
a localized hydration shell on the solute molecules. It took the
genius of Jean Perrin to settle this crucial question.

*Epilog I. The tortuous way of finding out.* In retrospect it
seems astounding that such a simple physical explanation of osmosis (and imbibition), while repeatedly advocated by outstand-
ing workers, could have been so thoroughly ignored. Negative
solvent pressure literally stares one in the face every time
an osmotic measurement is made either by push or pull technique
(Fig. 13). It would seem natural that *solvent tension* should
have been readily recognized so it could have served as a key
to understanding water relations in trees, but no, it took the
challenge of the old sap-rising problem to recognize water under
high negative pressure, and this was the Ariadne's thread that
finally led *us* to the crux of the matter: negative solvent pressure in free solutions!

## Brownian Motion and Jean Perrin: Solute and Solvent as Indepen-
dent Kinetic Units

We demonstrated by experiments on *matrices* and *suspensions* (81)
that the site of interaction with the mobile water at equilibrium
is the free surface, and only there. We have thus found that the
gravitational pressure gradients through matrices are that of

pure water, and indeed, if this were not so the Second Law would be violated. We also pointed out that there is no line of demarcation between a sedimented system and a self-dispersing suspension; we can change one into the other simply by changing the acceleration of gravity. The next question is obviously: What about the solute-solvent interaction or affinity dealing with colloidal and micro solutes? With focus on the phenomenon of Brownian motion, this question as everybody knows, evolved as one of the great landmarks in the development of the atomic theory.

Robert Brown (1773-1858, 46) was a Scottish botanist famous for a flora of Australia, discovery of the cellular nucleus and the Brownian motion. Quite incidental to taxonomic studies of fungi he observed that the spores suspended in water had a dancing movement when seen under the microscope. Dead particles of clay etc., proved to do the same, and even particles enclosed in water within a quartz crystal. He assumed, way ahead of his time, that this incessant motion must be imparted by the *thermal motion of the water molecules*, and it came to be known as the Brownian motion.

Ludwig Boltzmann at that time had already written his kinetic theroy of gases, including a formulation of the effect upon a diffuse gas when acted upon by a constant force like gravity. Kinetically he deduced the atmospheric formula which had already been established by another approach by Laplace: describing in exponential fashion the settling of the atmosphere in the gravitational field of the earth. Avogadro's number was known and appeared as a constant in Boltzmann's equation (p. 96, Eq.(28) where $m_2' = \frac{M}{N_A} (1 - \frac{\rho_1}{\rho_2}))$.

Jean Perrin (1909, 61), French physicist already famous for demonstrating the corpuscular nature of cathode rays, saw in the Brownian motion a possibility of furnishing experimental proof for the atomic theory, based essentially on the theorem of *equipartition of energy*.

"We have seen that the mean molecular energy is, at the same temperature, the same for all gases. This result remains valid when the gases are mixed. It is indeed known that each gas presses upon the enclosure *as if it alone were present*, that is to say that $n$ molecules of this gas develop in the volume $v$ the same partial pressure as if they were alone, in such a way that $\frac{3}{2} \frac{pv}{n}$ preserves the same value. On the other hand, when we try to repeat the reasoning which led to the relation $pv = \frac{2}{3} n w$, it is found that this reasoning remains applicable. Thus $w$ must preserve the same value. For example, the molecules of carbon dioxide and water vapor, present in the air, must have the same mean kinetic energy in spite of the difference of their natures and their masses."

And Perrin continues:

"This invariability of molecular energy is not confined to the gaseous state, and the beautiful work of Van't Hoff has established that it extends to the molecules of all dilute solutions. Let us imagine that a dilute solution is contained in a *semi-permeable* enclosure, which separates it from the pure solvent: we suppose this enclosure allows free passage to the molecules of the solvent, in consequence of which these molecules cannot develop any pressure (cf. Fig. 28A, p.39; our note), but that it stops the dissolved molecules. The impacts of these molecules against the enclosure will then develop an *osmotic pressure* P, and it is seen, if the reasoning is considered in detail, that the pressure produced by these impacts can be calculated as in the case of a gas, so that in consequence we write $Pv = \frac{2}{3} nW$, W signifying the mean kinetic energy of translation of n molecules contained in the volume v of the enclosure.

Now Van't Hoff has observed that the experiments of Pfeffer give for the osmotic pressure a value equal to the pressure which would be exerted by the same mass of dissolved substance if it alone occupied in the gaseous state the volume of the enclosure. W is thus equal to w: the molecules of a dissolved substance have the same mean energy as in the gaseous state."

Returning now to the Brownian motion, Perrin first derived a sedimentation equation based essentially on Boltzmann's theory of gases. From this it appeared that a uniform suspension would settle in conformity with his equation, provided the granules acquired the mean molecular energy of the water molecules. The concentration would then increase exponentially with depth as predicted and Avogadro's number could be calculated. So Perrin set about to prepare uniform emulsions of gum or resin by fractional centrifugations. He placed them in a sealed and thermo-regulated blood counting chamber and with sharp light flashes as illumination and a short focus microscope, he counted his way down the emulsion in 4 planes, 30 microns apart. Plugging the averages of many thousand counts into his sedimentation equation he found indeed Avogradro's number to average $7.05 \times 10^{23}$ as against a modern value of $6.05 \times 10^{23}$. Had the particles sedimented on the bottom or had they remained evenly suspended, the resulting values would have been respectively infinitely big or zero, but within this abyss he found his number within 17%!

Not only that, from his equation he could also predict the rate at which the granules would be displaced sideways. By an ingenious technique he tested this and found the same number; and measuring the random net rotation of relatively huge spheres (10-20 microns) as a function of time, was satisfied that even this checked!

He concludes:

"Thus the molecular theory of the Brownian movement can be regarded as experimentally established, and at the same time, *it becomes very difficult to deny the objective reality of molecules*. At the same time we see the law of gases, already extended by van't Hoff to dilute solutions, extended to uniform emulsions. The Brownian movement offers us, on a different scale, the faithful picture of the movements possessed, for example, by the molecules of oxygen dissolved in the water of a lake, which, encountering one another only rarely, change their direction and speed by virtue of their impacts with the molecules of the solvent.

It may be interesting to observe that the largest of the granules, for which I have found the laws of perfect gases apply, are already visible in sunlight under a strong lens. They behave as the molecules of a perfect gas, of which the gram-molecule would weigh *200,000 tons*."

Even the most diehard opponents of the atomic theory, one of whom was Wilhelm Ostwald, gave in although the great Ernst Mach did not (5, 61). Perrin emerged victor. The impact of his brilliant experimental work was enormous, and is the basis for *our* objections to the concept of the osmotic solute-solvent interaction being generated *within* a dilute solution at *equilibrium*. Nevertheless, interaction is, of course conspiciously displayed in the colligative properties of solutions: so if it is not generated *within* the solution, it is to the *boundaries* we must look!

*Epilog II. The Chemical Versus the Physical Mind*. The following quotation relating to the Perrin episode, is taken from the great book by Loeb and Adams, "The Development of Physical Thought" (46). It brings into focus the intriguing dichotomy between the chemical (thermodynamic) mind and the physical (kinetic) mind, and is not without parallel to our modest osmotic issue:

> "In the latter part of the nineteenth century, as a result of overemphasis on the purely thermodynamic treatment of problems, a group of scientists, chiefly in the field of chemistry, voiced the belief that it was a gratuitous assumption to speculate about the mechanism of the reactions and behavior of hypothetical atoms and molecules, in the face of the fact that such atoms and molecules had never been observed. The theory seemed to them particularly unnecessary and unfounded because one could in general calculate the results of most of the important physical phenomena on the basis of thermodynamic reasoning. The aggressive character of this school, headed as it was by a great teacher, Wilhelm Ostwald, caused it strongly to dominate the scientific thought of its time; and even though physicists generally accepted the kinetic theory, the development of the theory was somewhat retarded by the violent attacks made on it by the thermodynamicists...The suggestive analysis stimulated by the kinetic theory, however, enabled physicists not only to establish a clear kinetic concept of matter but also to discover the nature of the atom and the meaning of the periodic table, while the thermodynamic chemist was still mumbling that there was no atom,"

and, in the same vein today: there is no such thing as negative solvent pressure! (63)

## Further Theoretical Support for Solute-Solvent Coupling at the Free Surface

In 1937 Karl Herzfeld (40) wrote a classical paper entitled in translation: "Thermodynamic and kinetic views on the origin of vapor pressure lowering of solutions", where he develops the thesis that the lowering is caused by a solvent tension which is produced by reflection of hydrated solute molecules ($n_2$) from the free surface. As to ordinary solutions he states:

> "In this case the solute molecules or their hydration complex arrive at the solution surface and there are forced to turn around by forces between

solvent and complex, whereby they submit the surface to a well known kinetic force $\Pi = n_2\frac{RT}{V}$ . If we now compare the pure solvent at a pressure $P$ with the solution also at $P$ we see that the solvent molecules in this case can only give a force of $P - \Pi$ . When therefore one dissolves solute in solvent at an ambient pressure $P$, there is an apparent external pull $\Pi$ on the surface, which originates from the push $\Pi$ from within.

*If therefore one brings the solution in contact with solvent at pressure P, through a semipermeable membrane, then the solvent is pulled into the solution, because the free surface in contact with the air will yield to the kinetic pressure of the solute.* (our italics) The vapor lowering has therefore the same cause as the lowering through pull..."

"The evaporating molecules must perform work in order to force the solute to occupy lesser volume. This comes about because the vapor molecules lose velocity, i.e. kinetic energy, when they force the solute molecules to yield. Formally this parallels the velocity loss by reflection on a yielding mirror when the impact pressure is $\Pi$ ."

Herzfeld continues with a kinetic deliberation to explain the linear gravitational vapor pressure gradient transmitted in spite of the non-homogenous water structure at the surface.

Karol Mysels in his "Introduction to Colloid Chemistry" (1959, 55), independently develops the solvent tension theory, where momentum change of the solute molecules at the free surface constitute the driving force for solvent tension, and as a consequence the vapor pressure is lowered (Fig. 35).

Under the heading of "Meaning of osmotic pressure" he lucidly points out that a very simple kinetic picture is consistent with van't Hoff's gas analog.

"It is that particles of the solute move in the total volume of the solution as gas molecules move through their container. Whenever the particles encounter an impervious wall, they have to reverse direction and give up twice their momentum $mu$, just as gas molecules do at the walls of the container. When this reversal occurs at the limits of the solvent (be it an open surface or the wall of the vessel), it is the cohesive forces of the solvent that prevent the further travel of the solute particle and counteract the change of momentum. Thus the solvent is "stretched" or subject to an internally generated and negative pressure. As a result, the pressure exerted by the thermal motion of solute particles is not ordinarily perceived as such at the walls of a container of the solution, but vapor pressure of the solvent is lowered because of negative internal pressure to which it is subject."

"When a semipermeable membrane is encountered through which the solvent passes freely, it is no longer the solvent which limits the travel of the particles but only the membrane. Hence it is the membrane that must take up their change of momentum and is subject to a pressure. Figure 6-4 (our Fig. 35; our note) presents an equilibrium system of solvent, solution, and vapor, showing how the osmotic pressure and the solute is equivalent to a negative hydrostatic pressure exerted on the solvent both in maintaining osmotic equilibrium and lowering the vapor pressure."

Mysels also introduces "crowding pressure" as an explanation for the remarkably high osmotic effect of concentrated macromolecules.

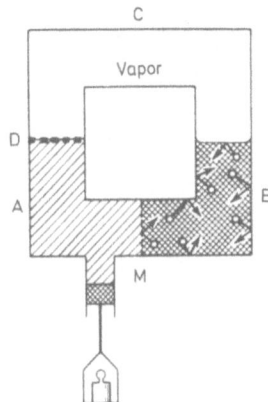

Fig. 35. Mysel's picture (6-4) of osmotic
equilibrium. *A*, solvent under tension, hanging on porous disk *D*. *B*, solution with solute molecules bombarding all boundaries.
*C*, equilibrium vapor pressure. *M*, semipermeable membrane

J. Duclaux (1965, 23), likewise sees osmosis as an expression
of negative solvent pressure, and rightly, draws a parallel between the "swelling" of a solution within a semipermeable bag
and the swelling of a gel:

"One must remember that the density of the solvent is not the same in the
solution as in its pure state. It is always less, and the solvent is in
fact dilated, i.e. under *negative* pressure. It is the sum of this *negative*
pressure of the solvent and the *positive* pressure of the solute particles,
which balance the external pressure. Seen in this way the osmotic pressure
is a real pressure, since there is no interaction between the dissolved
molecules and those of the solvent. In a dilute solution it (the osmotic
pressure) is the same as if the solvent did not exist; the water of hydration has no kinetic effect, for the hydrated solute acts as a kinetic
unit...As the solvent maintains a tension equal to the osmotic pressure,
it passes the membrane at a rate proportional to the tension and inversely proportional to the resistance of the semipermeable membrane, which
latter may be measured by the rate of filtration of the pure solvent."
(p. 15, sect. 3).

"When dialysis is performed against a pure solvent, the volume of the solution increases as the pure solvent seeks to equilibriate its partial
pressure through the membrane...We consider this movement as a swelling
of the solution by the solvent; it behaves like a lyophilic gel."
(cf. Fig. 11).

## Other Applications of the Kelvin-Poynting Column

We have repeatedly pointed out that a gel or suspension capable
of *transporting fluid water* must, at equilibrium, maintain a
normal gravitational pressure gradient in obedience to Pascal's
principle. Contrary to frequent claims (Boyer, 13; Slatyer, 93
etc.) there is no interaction with walls nor particles, i.e. no
"matric potential". The pressure effect on the mobile water is
strictly capillary and is located at the meniscus. This we have
already illustrated by means of several versions of the Kelvin-
Poynting column in Figs. 30-34. We shall add two more.

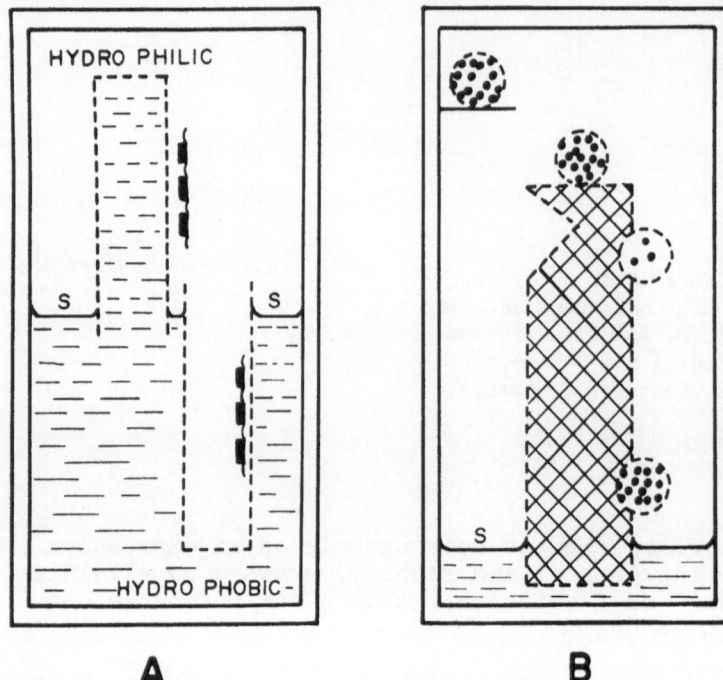

Fig. 36 A and B. (A) Callendar's isothermal "vapor sieve" columns for rapid
equilibration, inserts show details of menisci. (B) matrix column with peaks
and valleys and stiff semipermeable bags with solutions of any concentrations
at equilibrium. $S$ are the free surfaces

In Fig. 36A, is an immersed vapor column with hydrophobic pores
holding back convex water menisci. The gravitational increase
in vapor pressure from the free surface and down into the well
must match that emanating from the convex menisci (Poynting, 66;
Callendar, 15), not because of their curvature but because of
their increasing hydrostatic pressure. The same arguments, with
signs reversed, apply to the hydrophilic waterfilled column, left.

Similarly, fog drops do not acquire their increased vapor pressure from the curvature but from the increase in hydrostatic
pressure caused by the surface tension. It seems that this interpretation has escaped meteorologists (cf. Mordy, 54). Let us
consider a spherical fog drop originating from a salty condensation nucleus. If the surface tension of this drop balances
exactly the dispersal pressure of the solute molecules from
within, the solvent pressure becomes ambient. In an isothermal
enclosure this drop will transiently have the same vapor pressure as a flat water surface at the same level.

In Fig. 36, B is a matrix column with peaks and valleys, none
of which at equilibrium distort the regular gravitational attenuation in the vapor pressure emanating from the flat surfaces.
Nor, as already deduced by Gouy and Chaperon (31), do osmotic
cells, whether connected with the column or isolated from it,

depart from the normal hydrostatic gradient in the water. This formulation pertains very generally to plants in a state of no transpiration.

## D. Non Equilibrium States and the Role of Drag

We have so far dealt with the somewhat ephemeral concept of equilibrium states. Matrices, in general, offer no severe experimental problems in this respect, but solutions do, because of thermal convection (55, 61). We shall now relate the experimental study of two typical cases, namely diffusional and gravitational drag.

## Layering Water on the Surface of a Solution

Hulett (41) emphasized that when the diffusing solute molecules reach the free unrestrained solvent surface the dispersal pressure becomes balanced by an equal and opposite solvent pressure. We may now ask, what happens before they reach the surface? This has been studied by carefully layering water on top of a dextran solution held in an osmometer (81). It will be seen from Fig. 37 that during three hours there was a barely perceptible decrease in the osmotic pressure. By placing a printed grid behind the osmometer chamber it was also observed that the boundary layer between water and solution remained optically sharp for as long a time. This shows that the kinetic dispersal of the solute molecules is impeded by drag from the neighboring water molecules, which puts the solvent behind the advancing front under a tension, initially as great as before, i.e. equal and opposite to the solute pressure. The final state of equilibrium in Fig. 37 was obtained by mechanical mixing.

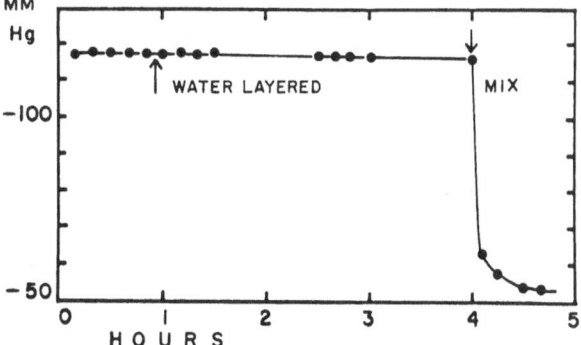

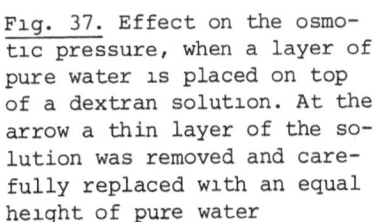

Fig. 37. Effect on the osmotic pressure, when a layer of pure water is placed on top of a dextran solution. At the arrow a thin layer of the solution was removed and carefully replaced with an equal height of pure water

## Hydrostatic Pressures in Stirred Suspensions

When a suspension of heavy particles is kept from sedimenting by stirring one would expect the hydrostatic pressure of the water itself would become that of the whole suspension. As this illustrates the buoyancy effect of regular solutes we shall

58

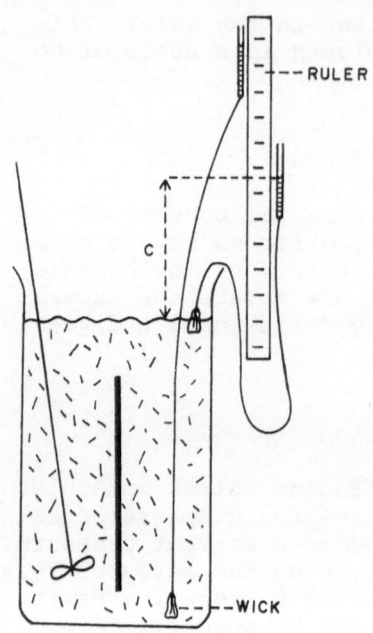

Fig. 38. Beaker of water, with emergy powder suspended by stirring. Baffle protects wick from direct propeller wash. $C$ is capillary blank from surface

give the results of experiments with three grades of emery powder, all of them washed and decanted until the supernatant remained clear. The densities of the suspensions were similar, respectively 1.275, 1.292 and 1.245. The hydrostatic pressure of the pure water was measured through a wick, placed alternately at the surface and at the bottom at a depth of 74 mm. Stirring was accomplished by a propeller, baffled to give a visibly uniform suspension with a reasonably flat surface (Fig. 38). If all granules were kept suspended by the turbulent water, their buoyant weight would appear as an increase in the hydrostatic pressure by drag. When measured at the bottom of our column the pressure would then rise above the free surface by $74 \cdot (1.275-1) = 20.4$ mm; respectively 21.6 and 18.1 in the two other suspensions.

The results are given in Fig. 39. It will be seen that the two finest suspensions gave the theoretical value for the buoyancy effect, i.e. a hydrostatic gradient equal to that of a fluid of the suspension density, something not attained by the coarser grit. The rate of settling is of course slower with finer grit. When sedimented, all sizes reacted to tapping by closer packing.[11]

---

[11] It may be noted that the so called turbidity currents in the ocean derive their increased density by turbulent suspension of fine sediment particles (Shepard, 1963, 92). A much debated spectacular incident happened in 1929 when an earthquake shook the Grand Banks. As a result several submarine cables broke one after the other, the furthest some 450 miles away. The spacing and timing of these breaks suggested a "landslide" and consequent turbidity current racing down the bottom slope, decelerating from 50 to 12 knots speed and showing a suggestive correlation with the bottom topography (Heezen and Ewing, 1952, 39).

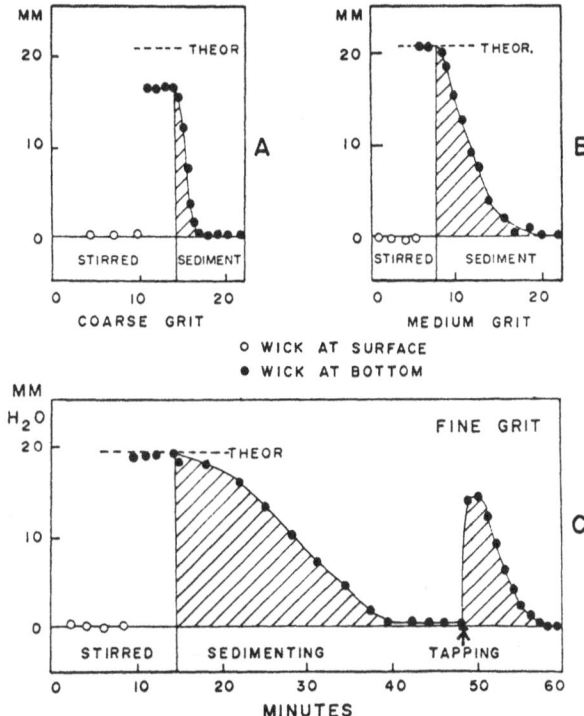

Fig. 39 A - C. Hydrostatic pressure of pure water in emery suspensions,
with and without stirring. (A) in a coarse suspension the stirring was unable to keep some of the heavy powder from settling. (B) and (C) in the
finer suspensions sedimentation took much longer time and all powder remained in suspension while stirring. With complete sedimentation the water
pressure returned to normal, but tapping the vessel compacted the sediment
with transient rise in hydrostatic pressure by drag. In the two finer suspensions the water pressure with stirring equals that of a fluid of the
density of the suspension, as indicated by the theoretical line

## E. Where is the Site of Coupling between Solute and Solvent?

Let us first discuss solute-solvent interaction under strict
equilibrium conditions. Perrin (61) showed that all interaction
*within* a solution between hydrated solute and solvent is random.
In Fig. 40A, a solution is resting on a semipermeable membrane
in an osmometer, and we read the osmotic pressure as the difference in level between the free water surface and the solution.
What causes this phenomenon?

One concept holds that the solute molecules lower the mol fraction of the water, which causes diffusion of water from the fully concentrated pure solvent below the membrane into the 'diluted'
solvent above. As ultimately the flux depends upon the concentration difference across the membrane, this places the site of interaction at the *membrane*. The other concept claims that the solvent pressure is lowered by solute impact against the *free surface*. Here then lies a possibility for an *experimentum crucis:*
where should we look, at the membrane or at the surface?

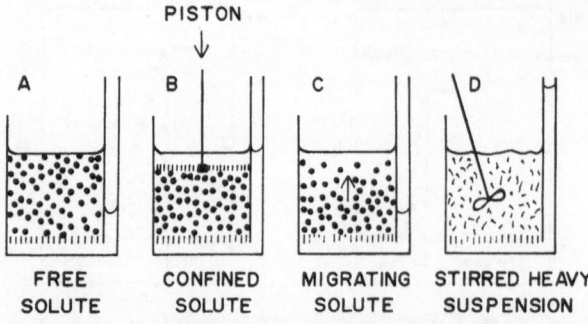

PISTON

| A | B | C | D |

| FREE SOLUTE | CONFINED SOLUTE | MIGRATING SOLUTE | STIRRED HEAVY SUSPENSION |

Fig. 40 A - D. (A) solute molecules strike free surface and put solvent under tension by simple force coupling. (B) solute molecules confined by a fixed semipermeable piston, returning the solvent to ambient pressure throughout. (C) water layered on solution, causing solute molecules to migrate upwards, putting solvent under tension as in A), but now by drag. (D) heavy suspension kept from sedimenting by stirring causes downward drag on water which acquires an increase in pressure equal to the negative buoyancy of the suspended particles

It is inherent in the solvent tension theory that *any factor* which can change the dispersal pressure of solute molecules besides *nRT* will change the colligative properties. Such effects can be induced by magnetic, electrostatic, gravitational and centrifugal forces, and they are striking in colloidal solutions through interaction (crowding) between the particles themselves. Thus, a change in size or charge of particles in high concentration will have a sensitive readout in the water tension. We shall describe below osmotic experiments involving crowding, solutes of positive and negative buoyancy and solutes susceptible to magnetic force.

## Effect of Crowding on Osmotic Pressure

One way to obtain an increase in the dispersal pressure of macro molecules is by crowding (55, 58, 81), so that finally the individual molecules through closer and closer contact attain essentially the elasticity of a matrix. That this is what happens is strongly suggested by the osmotic behavior of a series of dextrans with different molecular weights. The relatively enormous colligative changes of these polymers on a molal basis is well known (Fig. 41 left) even at high dilution. It is not so well known that they all very nearly approach the same magnitude of colligative properties (meaning simply negative solvent pressure) on a gram solute per cc basis irrespective of molecular weight. This means that when crowded together, or packed, by the negative solvent pressure the kinetics becomes drowned out by the mere space requirement of the monomer; it matters little whether the monomers are tied up in long or short chains (Fig. 41 right). Thus, when a colloidal suspension dehydrates, it greatly increases the dispersal pressure by mechanical crowding, approaching - if anything - Hooke's law of the spring, rather than the kinetic law of van't Hoff.

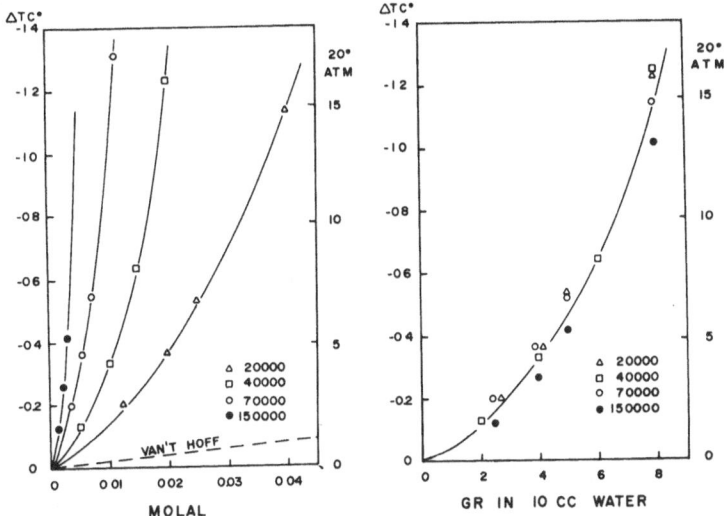

Fig. 41. Colligative properties in dextran solutions from 20,000 to 150,000 molecular weights, determined as melting points of microscopic ice crystals. *Left:* on a molal basis, *right:* on a gram per cm³ solvent basis. $\Delta T^{O}C$ = melting point depression. Dextran 20,000 has same concentration in either plot

Callendar (15) pointed out that the much too high osmotic pressure curves obtained by Berkeley and Hartley (10) in concentrated cane sugar solutions would follow van't Hoff's equation if one allowed some 5 moles of the water to be engaged in hydration of each mole of sugar. However, at concentrations as high as 500 g sugar per liter crowding undoubtedly is a factor, and such calculations are essentially a mathematical free-for-all unless supported by other physical evidence, like heat of hydration or otherwise.

## Effect of Gravity on Osmotic Pressure

For these experiments (90) a tall osmometer (Fig. 42) is filled to the brim with a solution of colloidal particles, either lighter or heavier than water. The free surface is closed by a slack saran film which makes it possible to turn the osmometer upside down, maintaining its null setting on the capillary. In so doing the buoyancy vector reverses with respect to membrane and free surface (Fig. 43).

Using a colloidal solute heavier than water the osmotic pressure increases when the osmometer is turned upside down, so that the free surface faces down. The increase corresponds closely to twice the buoyancy vector and remains constant whether the osmometer is artificially stirred (by a magnetic bar) or not (Fig. 44). This means that in the upright position the buoyancy vector diminishes the osmotic pressure by drag. If a true equilibrium could be established, half of the buoyancy difference would disappear when the solute particles became independently

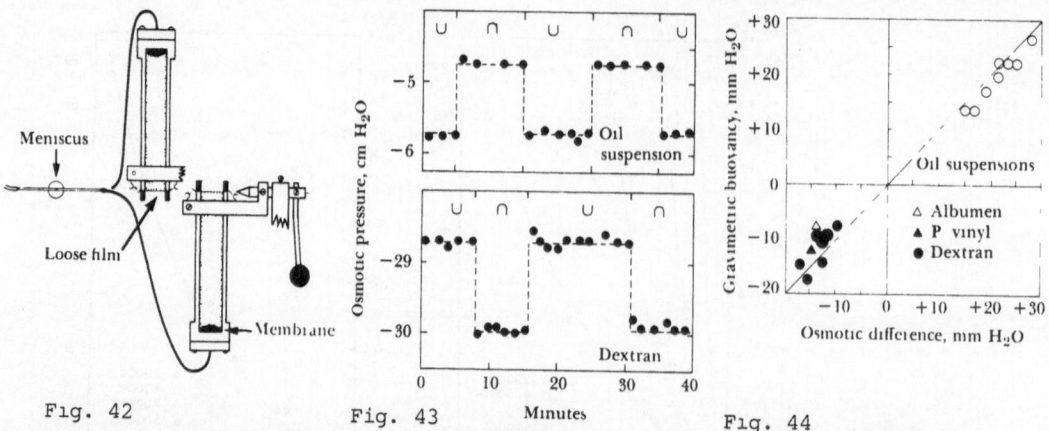

Fig. 42        Fig. 43        Fig. 44

**Fig. 42.** Pivoting osmometer, which when turned upside down, reverses the buoyancy vector of the solute in relation to the free surface and membrane. Slack film is clamped on surface to prevent water from falling out while still maintaining ambient pressure. Buoyancy changes in osmotic pressure correlate with free surface (90)

**Fig. 43.** Osmotic balancing pressures determined by the reversing osmometer in a heavier-than-water dextran solution and in a lighter-than-water oil suspension. U is upright position; ∩ is inverted position. When inverted, heavy dextran molecules increase the osmotic pressure; lighter oil particles decrease it. The increase correlates with the free surface, not with the membrane (82)

**Fig. 44.** Change in the osmotic pressure in the reversing osmometer as a function of the solute buoyancy determined gravimetrically, both expressed in mm $H_2O$. The relation is near 1/1. The increase is correlated with the free surface (82)

supported by the membrane, but this state is unattainable because of thermal convection.[12] With the free surface facing down, the buoyancy vector of the solute always adds to the osmotic pressure in full, whether the solute a) sinks through the convexing solvent or b) it is supported by the free surface, or both.

If instead of heavy solute molecules we use a light, stable oil suspension, like a machinist's "soluble" oil, we find again upon inverting the osmometer that the change in colloidal osmotic pressure is double the buoyancy effect and in this case the osmotic pressure is greater with the free surface at the top We see therefore that the interaction with the water correlates in either case positively and quantitatively with the solute

---

[12]Previously published data (60) using an insulated column of dextran 50 cm tall, were obtained by moving very slowly the membrane unit from top to bottom. We measured no clear decrease in osmotic pressure from sedimentation in the top position. This stands to be corrected: in recent checks with improved sensitivity the osmotic pressure was always less at the bottom, approaching the value for a stirred column.

pressure against the free surface, but negatively with the membrane. The cause of interaction is a simple force coupling at the free surface.

## Magnetoosmosis

An easier way to demonstrate the role of solute pressure against the free surface than the reversing osmometer is by means of magnetic solutions, such as manganese chloride. This compound has a very strong positive paramagnetic susceptibility, whereas water is weakly negative; on a molecular basis the ratio is near to +1000/-1. When a narrow cuvette of *saturated* $MnCl_2$ was slipped between the poles of a 6000 G magnet the solution surface rose locally 1.2 cm. This means that the magnetic force added to the thermal momentum of the solute and increased locally the huge (>100 atm) water tension by about 1 cm water (81).

Using instead a colloidal ferro magnetic fluid (Ferrofluidics Corp., Burlington, Mass.) quantitative measurements can be performed (89). The colloidal ferrite particles are thermally self-suspending and had, in our original preparation, a colloidal osmotic pressure of only a few cm $H_2O$. They were completely impervious to a dialyzing membrane. The magnet was slide-mounted above or below the osmometer so that the magnetic force could be kept constant, regardless of the changing distribution of the particles. The pressure effect on the water was measured with less than 1 mm$^3$ solvent displacements across the membrane, i.e. as a true null measurement. Only a diagram of principles and conclusions will be given here.

We see from Fig. 45 B that when the *magnet* is *above* the solution and the concentration of the ferrite increases toward the surface, the water tension *increases*. It was verified by different techniques (36, 89) that the magnetic force is additive to the thermal pressure. With the magnetic force kept constant the sum of the drag and of the momentum change at the surface (lift) becomes constant, and the *quantitative increase* in osmotic pressure proves that the site of action is at the free surface and nowhere else. Indeed, the difference in "water concentration" across the membrane is *lowered*, approaching theoretically zero.

With the *magnet below* the osmometer, i.e. with the force directed toward, or onto, the membrane, the "mol fraction difference" of the water across the membrane *increases* while the osmotic effect on the water *decreases*, or even turns negative by drag, and finally goes toward zero as the solutes vanish from the surface. Piling up on the membrane and pressing on it more and more, the solute molecules ultimately do nothing to the water.

We shall briefly describe new results obtained essentially with the same technique, but with a new ferro fluid greatly improved in strength and stability of colloidal dispersion. The osmometer (38) is attached to the end of a flat aluminum spring, and carries a wire extension to indicate its null position (Fig. 46).

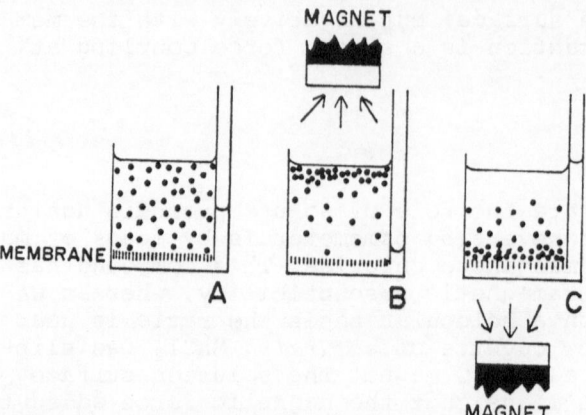

Fig. 45 A - C. Theoretical osmotic effect by colloidal ferro solution at equilibrium. (A) no magnet. (B) magnet above. (C) magnet below, normally convection adds some pressure in the capillary. *B* depicts a simple force coupling at the free surface

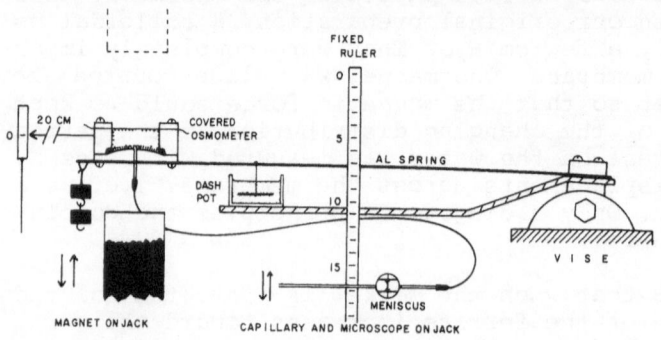

Fig. 46. Known vertical vector of magnetic force imposed on a colloidal ferro fluid in an osmometer. Negative pressure of the solvent is measured by the level of a movable capillary below its capillary blank value, at 0 on the fixed ruler

Five weights, each corresponding to $6.0^{13}$ mm water pressure in the osmometer can be hung at the end of the spring. The magnet (89) is mounted on a lab jack, either below or inverted above the osmometer. The fluid layer is 3-5 mm deep and is covered with a slack film.

With the *magnet* far *below*, or removed, the 5 weights were hung on the spring and the null indicator adjusted. One weight was removed and the magnet was elevated until the indicator was

---

[13]This number comes from the fact that the lead weights were originally made to hang centrally under the osmometer, and were calibrated to correspond to 5.0 mm change in water level of the osmometer. In the present experiments they were conveniently hung farther out on the spring, changing the calibration to 6.0 mm water.

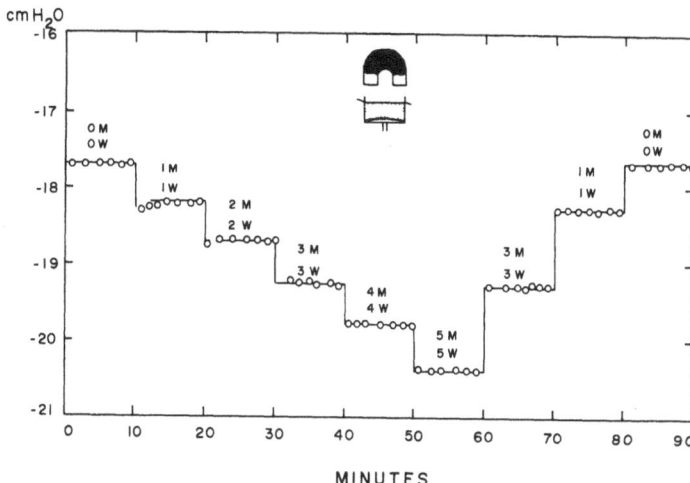

Fig. 47. Osmotic pressure in ferro fluid subjected to stepwise increase and decrease in magnetic force M, which is gauged by the number of equivalent weights W, hung on the osmometer (Fig. 46)

pulled down to the null position, where it was maintained. In so doing solvent flowed through the membrane unless checked by *elevating* the osmometer capillary by 6 mm. When the meniscus was observed through a microscope the equilibrium remained constant within 1 mm pressure for at least 10 min. Weights were removed, one by one, and as the new setting could be anticipated, equilibrium was quickly established. The sequence was then reversed, removing two weights at a time, back to the initial value.

With the *magnet above* the osmometer (Fig. 47) the ferrofluid must be tightly covered with a slack saran membrane. Null position was established without magnet and weights. Then one weight at a time was hung onto the spring and each time the magnet was lowered until the osmometer rose back to the null position. The read-out capillary must now be lowered 6 mm each time, and the equilibrium was checked for 10 min. The procedure was repeated until all five weights were hung, and then reversed as before.

Each of these two sequences required 90 min, and the six consecutive runs required three days in all. It will be seen from Fig. 48 that the curves, although not strictly at thermal equilibrium, show a remarkable degree of stability. This, undoubtedly, was a result of three factors: a weak magnetic force, a dominating thermal force, and a shallow layer of ferro fluid. A slight drift shows up when the first and last readings are compared, but is of no consequence to our main conclusions.

We know for the *magnet above* position that, if we diligently maintain a constant magnetic force by checking the null indicator, the total force on the water remains constant whether effected by drag or a new Boltzmann distribution, i.e. equilibrium or not, the magnetic force is quantitatively reflected by the

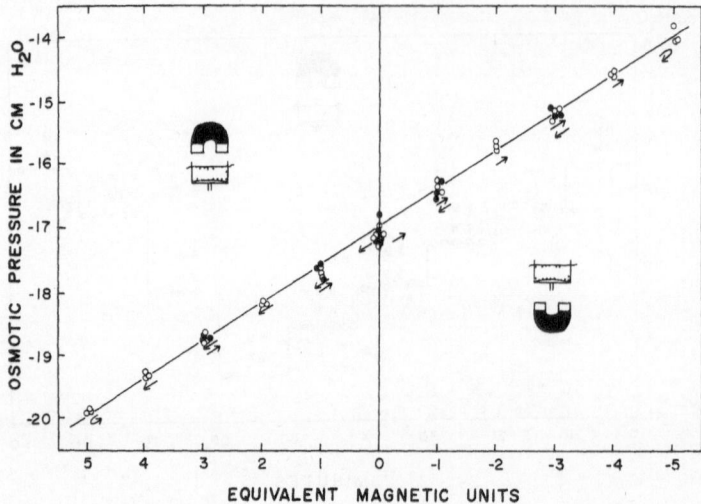

Fig. 48. Composite of 3 runs with magnet above and 3 with magnet below osmometer. Arrows indicate sequence in change of magnetic force as in Fig. 46.
*Diagonal* is the theoretical equivalence line between osmotic and magnetic
force where 1 M = 6 mm $H_2O$. Runs from the first two days have been lowered
in parallel onto the last run in order to compensate for a daily increase
(drift) in osmotic pressure of 5 mm $H_2O$

increase in the solvent tension. All our points do indeed fall
close to the equivalence line (Fig. 48). However, as most of
our points with *magnet below* also fall close to this line, it
means that a Boltzmann concentration is very weak in either
case.

The latter statement is also borne out by the fact that the magnetic force needed little, if any, adjustment to compensate for
drift. We therefore conclude that our *magnet below* experiments
can be extrapolated to an equilibrium condition where the osmotic force (= solvent tension) relates closely to the thermal
force *minus* the magnetic force. Above or below, at equilibrium
the site of coupling is exclusively the free surface; the membrane is transparent to water and has no effect on it.

## F. Answers to Part III

Lowered Vapor Pressure Over Solutions and Unsaturated Gels

We have demonstrated experimentally that the dispersal pressure
of solute molecules, or a gel matrix, against the free surface
at equilibrium, is balanced by a negative pressure (distension)
throughout the solvent. This, through Poynting's relation, gives
a full account of the lowering of the vapor pressure.

Osmosis by Bulk Flow

When a solution of macro molecules is separated from the pure
water by a semipermeable membrane the difference in hydrostatic
pressure of the water on either side will produce bulk flow when
the pore size is big enough, i.e. exceeds some 2 nm (59).


Equivalence of Osmotic and Hydraulic Flux

A 1 molar *solution* at 20°C kept in an open beaker measures 0 atm
(ambient) pressure. This represents a balance between two partial
pressures, a *solute pressure* of +24 atm and a *solvent pressure*
of -24 atm. Poured into an osmometer the solute molecules are
contained between the free surface and a semipermeable membrane
with pure water at ambient pressure below. Therefore, across
the pores there is an *hydraulic gradient* of 24 atm. The answer
to the equivalence is simple: both fluxes are hydraulic.


Osmotic Flux Against the Water Potential

A 1/100 molar solution of a macro solute A is separated from a
1 molar solution of a micro solute B, by means of a membrane
which excludes A but lets B through freely. Both solutions are
under ambient pressure. There is however, a hydrostatic difference in the *solvent* between the two solutions amounting to
-0.24 atm (A) and -24 atm (B). The *solution* B of ambient pressure can pass through the wide pores by bulk flow and is hence
drawn into A against the water potential, which in this case
is irrelevant. However, most membranes are imperfect; if a small
number of the pores exclude B, the flux becomes impeded. If a
great number of the pores are too small the flow may show an
initial reversal, i.e. solvent from A flows now largely by diffusion, into the much lower solvent pressure in B. But, because
solute B ultimately equalizes on both sides, the dispersal pressure of A takes over and the flux reverts from B to A. The mechanism is readily explainable by the solvent tension theory.


Osmotic Equilibrium

If the progress of the solute molecules is *stopped*, e.g. by a
rigid semipermeable membrane, by acceleration, by magnetic or
electrostatic force, or by any means which *prevents them from
reaching the free or unrestrained surface*, then the partial
pressure of the water itself returns to ambient throughout. For
instance, counter ions held by protein molecules in a Boltzmann
distribution have no individual osmotic effect, just as an hydration shell on a sugar molecule belongs osmotically to the
host. At equilibrium the solute-solvent coupling is generated
exclusively at the free surface. It is inherent in a Kelvin-
Poynting column of any kind at equilibrium that *the water is
only interrupted, not diluted by matrix or solutes, and that
its hydrostatic pressure is transmitted undiminished in all
directions according to the principle of Pascal.*

# V. Water Concentration Theory

The "water concentration theory" is the alternate, and so far only accepted way of attempting to explain the mechanism of osmosis. It seems inspired by the concept expressed for instance by Glasstone and Lewis (1960, 29):

"The osmotic pressure is brought into existence *only* when the solution is separated from the solvent by a semipermeable membrane. The resulting osmosis, or tendency for osmosis to occur, then produces an excess pressure in the solution."

In accord with this idea, the action of osmosis has been focused on solute-solvent interaction at the *membrane*. There seems to have been no firm theory set forth to explain the colligative properties.

Four worrisome points to be accounted for by any theory of osmosis have already been mentioned (p. 14): 1. Lowering of the vapor pressure over a solution, 2. Osmosis by bulk flow, 3. Equivalence of hydraulic and osmotic flux and 4. Osmotic bulk flow moving against the water potential.

A theme going through most of the recent literature is that the osmotic flux originates in a greater concentration (mol fraction) of water on the solvent side of the membrane than on the solution side, and this is somehow linked to the impact (momentum change) of the solute molecules against the pore openings. Pertaining to sap rising in trees we find a theory based on water interaction with standing gradients of solutes or matrices in the xylem vessel. We shall introduce the subjects by excerpts from the various authors and shall discuss their claims.

## Pressure Drop at the Pore Openings

Discussing osmotic equilibrium Pappenheimer, 1953 (59) says:

"In this case, the rate of diffusion caused by *the higher concentration* (our italics) (activity) of water on one side of the membrane containing no osmotically active solute molecules is exactly balanced by the increased thermodynamic activity of the water molecules subjected to hydrostatic pressure on the other side of the membrane."

In a similar vein Mauro (1957, 48) concedes:

"It should be emphasized that there is no kinetic theory in existence to explain the basis of the nondiffusional flux arising from the mole fraction effect. Unfortunately, the theory of liquids is inadequate at the present time for carrying out more than speculative analysis, but certainly as suggested by Lars Onsager, there must be a momentum deficiency in the microdomain of the pore in the solution side of the barrier. That is in a solid region of the barrier, the time average transfer of momentum is that prescribed by the hydrostatic pressure of the phase, but in the opening of the pore there is a deficiency since the *momentum arising from the macro molecule* (our italics) is not transferred to the solvent species in the pore, being cut off by the finite size of the pore. Thus, within the pore and only within the pore, a gradient of pressure arises and quasi laminar flow ensues from the solvent side to the solution."

Regarding the chemical potential profile at the pore openings of the membrane Mauro (1960, 49) states,

"Note that the decrement in chemical potential $\Delta\mu$ is equal to $RT \ln N_{H_2O}$ where a value of $N_{H_2O}$ different from unity has been established by the presence of the *absolutely* impermeant solute. It is apparent from a consideration of the interphase between the solution and the barrier that just inside the barrier, where there is only pure solvent, the continuity of the chemical potential function can only be satisfied by another component of the chemical potential, i.e. $\overline{V}\Delta p$. Thus, on the solution side of the interphase the decrement in chemical potential is $RT \ln N_{H_2O}$ and, in the barrier, $-V\Delta p$. It follows, then, that a drop of pressure, $-\Delta p$, must exist at the interphase of the barrier and the solution, and thus serves to act as the "driving force" for the osmotic flow."

"The argument pursued here is essentially similar to that advanced by Garby in 1957 (28) and might be classed as a thermodynamic approach. *Unfortunately, there is lacking at present an explicit kinetic theory which explains to origin of the pressure drop, and thus the osmotic flow.*" (our italics)

Let us define "origin" more precisely by stating that it includes both the *cause* and the *site* of the drop, neither of which could be defined.

Finally, having experimental evidence for negative solvent pressure Mauro (1965, 50) states,

"As shown by theory and confirmed by experiment, in the elementary osmotic system the pressure gradient causing flow of solvent through the membrane is established by a drop in pressure just within the membrane at the membrane solution interface. Moreover, the drop in pressure is not restricted to positive values but indeed can fall even to negative values, implying a tension state of the solvent within the barrier."

Peter Ray in 1960 (72) says on the subject:

"Available measurements on plant cells and tissues indicate that osmotic permeability exceeds that expected on the basis of diffusion by a factor of at least severalfold, which can not be explained by experimental errors. It thus appears, as previously observed with animal membranes, that osmosis is not strictly a diffusion process. The idea is examined that osmosis occurs by bulk flow of water through pores in the membrane. Occurrence of flow through pores appears to be an inevitable result of pressure gradients introduced within the pores by diffusion at pore apertures. This explains why an osmotic potential difference across a membrane causes as fast a water flow as a numerically equal hydrostatic pressure difference. The hypothesis can also be applied to osmosis by diffusible solutes, which is not proportional to differences in water deficit."

Jack Dainty (1963, 16) carried these ideas further and showed that a mathematical theory can be formulated on kinetic assumptions which fairly well fits the experimental evidence. Again,

the formulation is anchored in the concept of gradient in water
concentration at the pore openings of the membrane.

Ralph Slatyer (1967, 93) summarizes well the state of development of the water concentration theory:

> It is assumed, therefore, that within the pore, the pressure profile is
> exactly the same as in the pure water case and this indicates that mole
> fraction differences produce flow by the same type of mechanism as does
> a hydrostatic pressure difference, in both cases hydrostatic pressure
> gradients providing the operative forces. The reason for the sudden pres-
> sure change at the pore aperture has been pictured by both Ray (1960) and
> Dainty (1963) (cf. Fig. 5) as being due to the *greater concentration of
> water* (our italics) at the pore aperture than in the adjoining solution.
> Random activity of each water molecule results in a tendency to jump into
> any spaces which develop due to similar activity among its neighbors. Since
> there will be more spaces developed on the *less water-concentrated side*
> (our italics), more molecules jump from water to solution and the spaces
> so developed on the water side are filled by water molecules from further
> along the pore. The net creation of vacancies leads to a *decrease in den-
> sity* (our italics) of the water which can cause a decrease in pressure."

## Osmotic Flow Against the Water Potential

In dealing with solute mixtures or imperfect membranes, solute
separations and transient reverse flow is commonly observed.
Thermodynamic *reflection coefficients* have been introduced by
Staverman (1951, 94) to account for such events. Thermodynamics
also lends itself to treatment of other events, involving chemical potential of all species, as carried out by Kedem and
Katchalsky (1958, 42) or by Mauro's procedure (1960, 49) to establish "the thermodynamic profiles within the barrier and thus
to predict the movement of the various species...".

## Anti-Gravity Devices in Trees

Regarding measurements of the negative pressure in the vascular
system of plants including that obtained with the pressure chamber, Slatyer (93) states:

> "Such measurements combined with measurements of the osmotic pressure in
> the xylem sap...should give an adequate picture of the water potential in
> continuous liquid-filled columns which are under tension. It should be re-
> membered, though, that much of the water in vascular tissue, as the water
> potential falls, may be retained by matric forces rather than by cohesion."

Plumb and Bridgman (64) derive from thermodynamic analysis that
the hydrostatic gravitational pressure gradient in a vertical
column can be abolished by providing it with a "constrained
chemical activity gradient", to wit, a concentration gradient
of either a gel or attached wiggling molecular hairs. Levitt
and Storvick (43) challenge this by pointing out that neither
hairs nor gels have been observed in xylem vessels. Furthermore,
the freezing point of xylem sap *in situ* in evergreens is near
that of pure water and is therefore not depressed by a gel or
attached hairs (Hammel, 35a).

## Comments

Let us frankly admit that we are unable to visualize the mechanical aspects of the water concentration theory. We shall therefore limit ourselves to comment on some seemingly untenable thermodynamic consequences of the proposed theory.

The hydrostatic pressure of a *solution* kept in an open beaker is ambient, and the water concentration theory holds that the solvent pressure is also ambient; but for "some reason connected with the presence of solute molecules" (29) there is a decrease in the chemical potential which lowers the vapor pressure. Likewise, if the solution is kept submerged within a semipermeable enclosure (Fig. 49B) the hydrostatic pressure of the *solution* will be elevated and therefore also that of the solvent; but again the presence of the solute molecules, somehow, lowers the vapor pressure to match that of the outside water.

If it were true, however, that the hydrostatic pressure of the mobile water (which this is all about) were changed one way or the other *by the mere presence* of solute molecules, attached molecular hairs (64), or a matrix (93), then the *densitiy of the compressible water* would change. This in itself would suffice to maintain a *perpetual flow* a) if a stiff matrix were immersed in water, b) if a solution at equilibrium within a semipermeable cylinder were immersed in water (Fig. 49, A and B), or c) if even a simple capillary were immersed in water; that is, unless one were also to invoke an inadmissable change in the mass of the water.

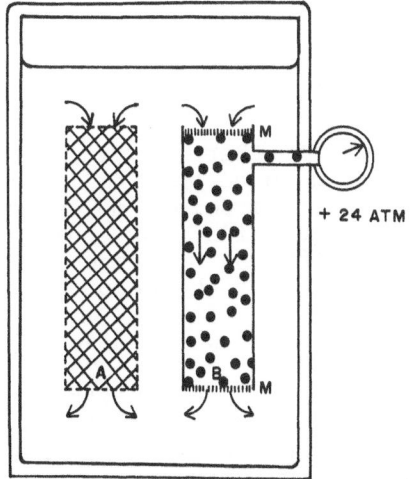

Fig. 49. *A*, saturated matrix; *B*, a 1 Mol solution within a semipermeable cylinder, both submerged in isothermal water. When at equilibrium, outside and inside pressure and hence density of the mobile water in *A* and *B* must match at each level, otherwise perpetual flow would ensue, as indicated by the arrows. *M* is membrane

As to the proposed action at the pore openings of the membrane (Fig. 50), why does not the "molfraction of the chemical potential" extend into the fine pores of the membrane, where the "water concentration": is, in fact, even less than in the solution?

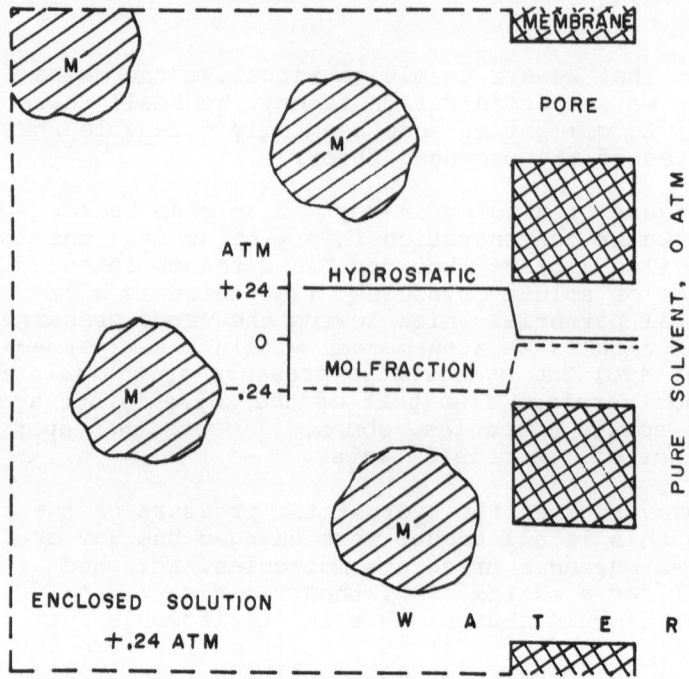

Fig. 50. Pressure profile through a semipermeable membrane, invoking water concentration theory. On the left of the stiff membrane is a closed container holding a 0.1 Mol ideal solution of macro-molecules M. At equilibrium the solution pressure accordingly measures +0.24 atm at 20°C. The pressure of the solvent is assumed to be the same but its chemical potential is assumed to be raised to zero (ambient) by a molfraction effect on the water. Steep (0.24 atm) gradients cancel each other in the pore openings. To the right of the membrane is pure water. Gravity is zero throughout (compare Fig. 29A)

We may further ask: in a dilute solution, *at a given moment*, only one pore opening in a thousand may be struck by, or interact with, a solute molecule, while all the rest see pure water on either side. How could an hydraulic solvent gradient be maintained by the membrane in face of such perpetual back-leaks, in violation of the Second Law?

If the pores were 10 nm across and the membrane 100 μm thick the pores would be 1000 times longer than wide. Would not the combined solute impacts result in an elastic compression of the rigidly supported membrane, and a direct hit on a pore opening result in a reversible pressure pulse, (if any), rather than a one-way pumping action? As for the *de facto* osmotic-hydraulic *equivalence* we are given only a circular assertion that the solute interaction at the pore openings would accomplish just that.

The reflection coefficient explains how solute and solvent may move *together* but not *what it is* that moves them; and a full thermodynamic inventory may predict flow against the water potential but indicates nothing about the driving force.

The *solvent tension theory*, such as illustrated in Fig. 13, avoids all of these stumbling blocks, and suggests that the solvent pressure in Fig. 50 is ambient throughout: in the outside solvent, through the membrane and within the solution. The pressure of the solution against the membrane is caused by the solute molecules alone.

The profile in Fig. 50, embodying the *water concentration theory*, amounts to a *fictitious* positive solvent pressure cancelling a *fictitious* negative molfraction effect.


## Concluding Remarks


The lowering of the chemical potential of water by a solute in an open solution is caused solely by the lowering of the hydrostatic pressure of the water. The elevated hydrostatic pressure of a *solution* within a stiff semipermeable enclosure submerged in water is caused solely by the thermal dispersal pressure of the solute; the water pressure is ambient throughout, the "chemical potential" of the water remains zero, and the assumed chemical interaction within the water is fictitious, as shown by Perrin, the Kelvin-Poynting column, by magnetic and gravitational experiments. In regard to osmosis and imbibition an irrelevant chemical nomenclature has led to much confusion, and might as well be dropped in favor of the physically meaningful term *water tension*, so much the more so, as we shall show in Chapter III that water tension is the sole cause of *all* colligative properties of a solution.

# Chapter II  Some Reversible Thermodynamic Relationships at Equilibrium

We shall review the derivation of two important thermodynamic relationships required for the next Chapter. These are the Poynting relation and the Clapeyron equation. They are required to derive the effects of temperature and hydrostatic pressure upon the vapor pressure of a liquid or its solid phase. From the relationships between vapor pressure and hydrostatic pressure applied to the liquid and its solid, and the relationships between vapor pressure and temperature of the liquid and its solid, we shall then derive the effects of hydrostatic pressure on melting and boiling temperatures in Chapter III. In Chapter II, we shall also review the application of thermodynamic principles to the treatment of a solution of solute in a solvent and then discuss the limitations of thermodynamic statements for interpreting the basic effect of the solute upon the solvent. This will lead to a discussion in Chapter IV of the alternatives for elucidating the essential features of solute-solvent interaction which induce the colligative properties of real solutions. For a more thorough review of chemical thermodynamics, the student should refer to current textbooks on the subject.

This review of thermodynamic statements is intended only for students not familiar with the language of thermodynamics or with its basic principles and use. Although the colligative properties of a solution are attributable to the kinetic and mechanical behavior of solvent and solute molecules, ions, particles, etc., the relationships between the colligative properties of a solution are most efficiently expressed in thermodynamic terms.

## Some Basic Thermodynamic Statements

In accordance with the First Law of Thermodynamics, a small amount of heat $\delta Q$ entering a system from its environment minus a small amount of work $\delta W$ performed by the system upon the environment results in a small change $dE$ in the internal energy of the system, i.e.

$$dE = \delta Q - \delta W.$$

Both $\delta Q$ and $\delta W$ depend on the path by which heat enters the system and on the path by which work is performed by the system and are therefore designated as inexact differentials. Energy is a function of the state of the system so that $dE$ is an exact differential and depends only on the state of the system before and after the change in energy. This equation is valid for all processes, reversible or not. Another function of the state of a system is the entropy $S$ of the system so that $dS$ is also an exact differential. As an outgrowth of the Second Law of Thermo-dynamics, $dS$ is defined as $\delta Q/T$ for a reversible process and

$\delta Q = TdS.$

Considering a system for which all work is pressure-volume work so that $\delta W = pdV$ for reversible processes, then $dE$ becomes

$dE = TdS - pdV.$

This equation is also valid for irreversible processes, since for these processes $\delta Q > TdS$ and $\delta W > pdV$ by the same amount so that the difference between $TdS$ and $pdV$ is unaffected by the inequalities provided $T$ and $p$ are equilibrium values.

Enthalpy is defined as $H = E + pV$ and the Gibbs free energy is defined as

$G = H - TS$

so that

$G = E + pV - TS$

and

$dG = dE + pdV + Vdp - TdS - SdT.$

Substituting for $dE$ from above then,

$dG = -SdT + Vdp.$

The above relationships were developed for constant composition and mass. However, the Gibbs function is an extensive function of the state of the system and depends upon the number of moles and species of molecules $n_i$, as well as upon temperature and pressure, i.e. $G$ $(T, p, n_1, n_2, \ldots)$. If the $n_i$'s vary then the differential of the Gibbs function has additional terms, so that

$$dG\ (T,p,n_1,n_2\ldots.) = \left(\frac{\partial G}{\partial T}\right)_{p,n_1,n_2..} dT + \left(\frac{\partial G}{\partial p}\right)_{T,n_1,n_2..} dp$$

$$+ \left(\frac{\partial G}{\partial n_1}\right)_{T,p,n_2,..} dn_1 + \left(\frac{\partial G}{\partial n_2}\right)_{T,p,n_1,n_3,..} dn_2 + \ldots$$

Since $dG = -SdT + Vdp$ for constant $n_i$, then

$$dG\ (T,p,n_i) = -SdT + Vdp + \left(\frac{\partial G}{\partial n_1}\right)_{T,p,n_2,n_3..} dn_1$$

$$+ \left(\frac{\partial G}{\partial n_2}\right)_{T,p,n_1,n_3,..} dn_2 + \ldots$$

Like $G$, $V$ is also an extensive function and a function of $T$, $p$, $n_1$, $n_2$,.. For these functions, partial molar quantities can be defined for each as follows:

$$\left(\frac{\partial V}{\partial n_i}\right)_{T,p,n_1,n_2,\,..} = \bar{V}_i$$

and

$$\left(\frac{\partial G}{\partial n_i}\right)_{T,p,n_1,n_2,\,..} = \bar{G}_i = \mu_i \; .$$

The partial molar Gibbs free energy is also designated as the chemical potential, $\mu_i$, of the $i$th component of the system. Thus

$$dG = -SdT + Vdp + \mu_1 dn_1 + \mu_2 dn_2 + \; .\,.\,.\,.$$

For an isothermal process, the rate of change of the Gibbs function with pressure is

$$\left(\frac{\partial G}{\partial p}\right)_{T,n_1,n_2,\,..} = V \; .$$

By differentiating this relationship again with respect to $n_i$ and employing the terms for the partial molar functions

$$\left(\frac{\partial \frac{\partial G}{\partial n_i}}{\partial p}\right)_{T,n_1,n_2,\,..} = \left(\frac{\partial V}{\partial n_i}\right)_{T,n_1,n_2,\,..}$$

then

$$\left(\frac{\partial \mu_i}{\partial p}\right)_{T,n_1,n_2,\,..} = \bar{V}_i \; .$$

These thermodynamic relations apply without regard for the number of species of molecules in the system, the equation of state for the system or the phase of the system. Since our particular interests are with the changes that occur in solvent when other molecules are added to the solvent, these fundamental thermodynamic relations shall be applied only to solvent and solutions.

In the next two sections of this Chapter we shall be discussing a single species of molecule, the solvent, for which $n_2$, $n_3$,... are zero and $n_1$ is the number of moles of solvent molecules. In this system, a change in the free energy is expressed by the thermodynamic statement

$$dG = -SdT + Vdp + \mu_1 dn_1$$

or a change in the molar free energy is expressed by

$$d\bar{G} = -\bar{S}dT + \bar{V}dp .$$

## The Poynting Relation

Let us consider what the effects on pure solvent are when changing the hydrostatic pressure applied to the liquid only.

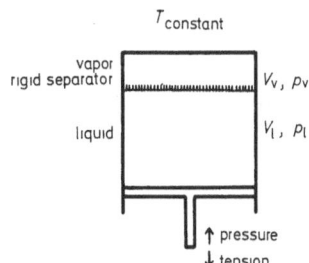

Fig. 51. A system to illustrate the effects of positive pressure or negative pressure (tension) applied to a liquid, $p_\ell$, upon its vapor pressure, $p_v$. The rigid separator passes only vapor and adheres to the liquid. The gravity field is, $g = 0$. $V_\ell$ and $V_v$ are the volumes of the liquid and the vapor respectively

Fig. 51 illustrates how the effects may be observed. Liquid is separated from the vapor by a rigid separator which passes vapor freely but does not pass liquid. The vapor is enclosed in a fixed volume $V_v$ above the separator and the liquid is enclosed below the separator in volume $V_\ell$ by a piston in a cylinder. The liquid is nearly incompressible so that $V_\ell$ may be considered constant for low pressures. The vapor pressure, $p_v$, will be variable as the liquid pressure $p_\ell$, varies. If the pressure $p_\ell$ is caused to increase slowly so that equilibrium is preserved between the liquid and the vapor, then, as we shall show in the next Section, during vaporization the change in the molar free energy will be the same in the liquid and in the vapor, i.e.

$$\int d\mu_\ell = \int d\mu_v$$

If the process is isothermal, then

$$\int \bar{V}_\ell dp_\ell = \int \bar{V}_v dp_v \tag{1}$$

which is known as the Poynting relation.

The Clapeyron Equation

First, we need to consider the change in free energy accompany-
ing the isothermal, isopiestic (constant temperature and constant
pressure) conversion of liquid to vapor (and solid to vapor).
From the First Law of Thermodynamics

$$\Delta E = Q - W .$$

Considering a system for which all work is pressure-volume work
at constant pressure,

$$Q_p = \Delta E + p\Delta V .$$

The enthalpy is defined as a function such that its change at
constant pressure will equal $Q_p$, i.e.

$$H = E + pV$$

and

$$\Delta H = \Delta E + p\Delta V = Q_p .$$

From the Second Law

$$Q = T\Delta S ;$$

so that

$$\Delta H = T\Delta S$$

at constant pressure.

If the process of vaporization is carried out at the equilibrium
pressure, then, from this relationship the entropy change of vaporization, $\Delta S_{vap.}$, will be

$$\Delta S_{vap.} = \frac{\Delta H_{vap.}}{T}$$

Since the free energy change for any isothermal process is, by
definition,

$$\Delta G = \Delta H - T\Delta S$$

we can conclude for vaporization from liquid or solid that

$$\Delta G = 0.$$

In general, this is true for any transition between phases at
equilibrium pressure.

The Poynting relation was based on the recognition that the free
energy change for any phase transition must equal zero at equilibrium. We can also determine how the vapor pressure must vary

with temperature so that $\Delta\bar{G} = 0$. For the vapor phase

$$d\bar{G}_V = -\bar{S}_V dT + \bar{V}_V dp_V$$

and for the liquid phase

$$d\bar{G}_\ell = -\bar{S}_\ell dT + \bar{V}_\ell dp_\ell$$

where the temperature is the same for both phases while $p_\ell$ and $p_V$ may differ. The total differential of the change in molar free energy for the evaporation process is the second equation subtracted from the first

$$d(\bar{G}_V - \bar{G}_\ell) = -(\bar{S}_V - \bar{S}_\ell)dT + \bar{V}_V dp_V - \bar{V}_\ell dp_\ell$$

or

$$d\Delta\bar{G} = -\Delta\bar{S}dT + \bar{V}_V dp_V - \bar{V}_\ell dp_\ell$$

if equilibrium is maintained before and after the change. Since $\Delta\bar{G} = 0$ under the circumstances and since $\Delta\bar{S} = \dfrac{\Delta\bar{H}}{T}$, we have

$$-\frac{\Delta\bar{H}}{T}dT + \bar{V}_V dp_V - V_\ell dp_\ell = 0 \ . \tag{2}$$

If the pressure on the liquid is constant, then

$$\left(\frac{\partial p_V}{\partial T}\right)_{p_\ell} = \frac{\Delta\bar{H}_{\ell V}}{T\bar{V}_V}$$

where $\Delta\bar{H}_{\ell V}$ is the molar enthalpy for transition from liquid to vapor and $\bar{V}_V$ is the molar volume. If the change in pressure on the liquid and vapor are the same, then we may write from Eq.(2)

$$\frac{dp_V}{dT} = \frac{\Delta\bar{H}_{\ell V}}{T(\bar{V}_V - \bar{V}_\ell)} \tag{3}$$

which is the Clapeyron equation.

## Application of Thermodynamic Statements to Solutions

In our derivation of the Poynting relation and the Clapeyron equation, we considered these equations as they apply to a single species of molecule, i.e. the molecules were the same in all three phases of the substance. In this treatment we relied on the important thermodynamic statement that a change in the Gibbs free energy in a fixed amount of substance is given by

$$dG = -SdT + Vdp \ .$$

The Gibbs free energy, $G(T,p,n_i)$, the entropy, $S(T,p,n_i)$ and the volume $V(T,p,n_i)$ are all extensive functions of the system, that is they are all homogeneous functions of the number and species of molecules, $n_i$, in the first degree. They are also homogeneous functions of temperature and pressure in the first degree. Euler's theorem states that for a completely general homogeneous function of degree $\alpha$ in the variables $x$, $y$, $z$, ....,

$$f(\lambda x, \lambda y, \lambda z, \ldots) = \lambda^{\alpha} f(x,y,z,\ldots);$$

and differentiating with respect to $\lambda$,

$$\frac{df(\lambda x, \lambda y, \ldots)}{d\lambda} = \frac{\partial f}{\partial \lambda x} \frac{\partial \lambda x}{\partial \lambda} + \frac{\partial f}{\partial \lambda y} \frac{\partial \lambda y}{\partial \lambda} + \ldots$$

$$= x\frac{\partial f}{\partial \lambda x} + y\frac{\partial f}{\partial \lambda y} + \ldots$$

$$= \alpha \lambda^{\alpha-1} f(x,y,\ldots) . \tag{4}$$

For all extensive functions, multiplying the number of moles of each species by a factor $\lambda$ is the same as multiplying the extensive function by the same factor $\lambda$, e.g.

$$G(T,p,\lambda n_1, \lambda n_2, \ldots) = \lambda G(T,p,n_1,n_2,\ldots) .$$

Application of Euler's theorem, Eq.(4), to the special case for $\lambda = 1$ leads to an important characteristic of all extensive functions e.g.

$$V = n_1\left(\frac{\partial V}{\partial n_1}\right)_{T,p,n_2,\ldots} + n_2\left(\frac{\partial V}{\partial n_2}\right)_{T,p,n_1,\ldots} + \ldots$$

$$S = n_1\left(\frac{\partial S}{\partial n_1}\right)_{T,p,n_2,\ldots} + n_2\left(\frac{\partial S}{\partial n_2}\right)_{T,p,n_1,\ldots} + \ldots$$

$$G = n_1\left(\frac{\partial G}{\partial n_1}\right)_{T,p,n_2,\ldots} + n_2\left(\frac{\partial G}{\partial n_2}\right)_{T,p,n_1,\ldots} + \ldots$$

Partial molar functions for these and other extensive functions have already been defined on p. 76, for example,

$$\left(\frac{\partial V}{\partial n_i}\right)_{T,p,n_j} = \overline{V}_i$$

$$\left(\frac{\partial S}{\partial n_i}\right)_{T,p,n_j} = \overline{S}i$$

$$\left(\frac{\partial G}{\partial n_i}\right)_{T,p,n_j} = \overline{G}_i \equiv \mu_i \ .$$

The partial molar Gibbs free energy of the $i$th component, $\overline{G}_i$, is of such special importance that it has been designated as the chemical potential $\mu_i$ of the $i$th component.

All partial molar quantities are intensive functions of the $i$th component, that is, they are homogeneous functions in zero degree in $n_i$, e.g.

$$\overline{G}_i(T,p,\lambda n_1,\lambda n_2,..) = \lambda^0 \overline{G}_i(T,p,n_1,n_2,..) \ .$$

Increasing the number of moles of each species by a factor $\lambda$ has no effect on the intensive function. These intensive functions are, however, homogeneous functions in the first degree of temperature and pressure. Since $\overline{G}$ is a homogeneous function in degree zero of $n$, applying Euler's theorem to $\overline{G}$, for example, gives

$$n_1 \left(\frac{\partial \overline{G}_2}{\partial n_1}\right)_{T,p,n_2,..} + n_2 \left(\frac{\partial \overline{G}_2}{\partial n_2}\right)_{T,p,n_1,..} +... = 0 \ .$$

Since

$$\frac{\partial \overline{G}_1}{\partial n_2} = \frac{\partial^2 G}{\partial n_2 \partial n_1} = \frac{\partial \overline{G}_2}{\partial n_1} \ ,$$

then

$$n_1 \left(\frac{\partial \overline{G}_1}{\partial n_2}\right)_{T,p,n_1,..} + n_2 \left(\frac{\partial G_2}{\partial n_2}\right)_{T,p,n_1,..} +... = 0 \ .$$

When applied to a two component system, this relationship is known as the Gibbs-Duhem equation. It tells us, for example, that if the amount of solvent, $n_1$, is held constant, then an increase in the chemical potential of the solute by addition of solute must result in a decrease in the chemical potential of the solvent according to the relation

$$\frac{\partial \mu_1}{\partial n_2} = -\frac{n_2}{n_1}\frac{\partial \mu_2}{\partial n_2} \ .$$

Like the other partial molar quantities, the chemical potential of the $i$th component is a function of the temperature and pressure in the first degree and a function of $n_i$ in zero degree. Thus changing the $n_i$'s by the same amount $\lambda$ will not change these intensive quantities. However, changing the ratios of the $n_i$'s will change the chemical potential of the $i$th component. Likewise the chemical potential depends upon the mole fraction of the components subject to the condition that the sum of the mole fractions equal unity. Thus we may write the partial molar free energy as a homogeneous function in the first degree by expressing its dependence upon the mole fractions of the several

components, $x_i = \dfrac{n_i}{\sum\limits_j n_i}$ . Expressed this way, then, changing the

mole fraction of all components by a factor $\lambda$ will have the same effect upon the chemical potential of the $i$th component as multiplying the chemical potential by $\lambda$ e.g.

$$\overline{G}_i(T,p,\lambda x_1,\lambda x_2,\ldots\lambda x_{j-1}) = \lambda \overline{G}_i(T,p,x_1,x_2,\ldots x_{j-1})$$

where the number of independent $x_i$'s is one less than the number of components $j$ and where $\sum\limits_{j-1} x_i = 1$. Since the chemical potential

$$\mu_i = \left(\frac{\partial G}{\partial n_i}\right)_{T,p,n_{i},\ldots} = \overline{G}_i \ ,$$

is an intensive quantity which can nevertheless be expressed as a homogeneous function in the first degree in the mole fraction of its components, similar relationships may be found for it as were found for $G$. Since $\mu_i$ is a function of $T$ and $p$ as well as $x_i$, its total differential will be

$$d\mu_i = \left(\frac{\partial \mu_i}{\partial T}\right)_{p,x_1,x_2,\ldots} dT + \left(\frac{\partial \mu_i}{\partial p}\right)_{T,x_1,x_2,\ldots} dp + \left(\frac{\partial \mu_i}{\partial x_1}\right)_{T,p,x_2,\ldots} dx_1$$

$$+\ldots \qquad + \left(\frac{\partial \mu_i}{\partial x_{j-1}}\right)_{T,p,x_1,x_2,\ldots} dx_{j-1} \ .$$

We note that

$$\frac{\partial \mu_i}{\partial T} = \frac{\partial^2 G}{\partial T \partial n_i} = -\frac{\partial S}{\partial n_i} = -\overline{S}_i$$

and

$$\frac{\partial \mu_i}{\partial p} = \frac{\partial^2 G}{\partial p \partial n_i} = \frac{\partial V}{\partial n_i} = \overline{V}_i$$

thus

$$d\mu_i = -\overline{S}_i dT + \overline{V}_i dp + \left(\frac{\partial \mu_i}{\partial x_1}\right)_{T,p,x_2,..} dx_1$$

$$+ ... \qquad + \left(\frac{\partial \mu_i}{\partial x_{j-1}}\right)_{T,p,x_1,..} dx_{j-1} \; .$$

This equation has become important for providing a thermodynamic description of the change in chemical potential in a system involving a solvent and solutes. For example, if the mole fraction of a single solute is $x_2$ and the mole fraction of the solvent is $x_1$ then the change in chemical potential of the solvent, $d\mu_1$, caused by a change in the mole fraction of solute, $dx_2$, becomes

$$d\mu_1 = -\overline{S}_1 dT + \overline{V}_1 dp + \frac{\partial \mu_1}{\partial x_2} dx_2 \; . \tag{5}$$

It is understood here that *the change in pressure, dp, refers only to pressures applied externally to a homogeneous solution.*


Ideal Solution

In the next Chapter, we shall turn to a simplified model of a solution in order to treat the kinetic phenomena which may account for the colligative properties of a solution. We shall begin by treating the ideal solution, i.e. where the volume equals the sum of its components. To understand the meaning of the ideal solution, we consider again the fact that the volume of a solution may be determined by the expression:

$$V = n_1 \left(\frac{\partial V}{\partial n_1}\right)_{T,p,n_2,..} + n_2 \left(\frac{\partial V}{\partial n_2}\right)_{T,p,n_1,..} + ...$$

We recall that this expression was derived by an application of Euler's theorem to the extensive function $V(T,p,n_1,n_2,...)$ which is homogeneous in the first degree in $n_1,n_2,....$. Although a precise statement, this expression in no way reveals how we might predict what the volume of a solution of solvent and solutes might be. We must resort to empirical determinations. However, we can anticipate that if adhesive forces between unlike molecules are greater than cohesive forces between like molecules, then shrinkage will occur upon mixing and the partial molar volumes will be less than the molar volumes of the pure components. If the intermolecular forces are weaker between unlike molecules than between like molecules then expansion will occur and the partial molar volumes will be greater than the molar volumes of the pure components. Only if the interactions between like and unlike molecules are the same, will the partial molar volumes be equal to the pure molar volumes at all concentrations so that the volume per mole of solution equals $\sum_i x_i \overline{V}_{pure\;i}$. Such a solution is an ideal solution.

## Interpreting the Change in Chemical Potential of a Solvent in a Homogeneous Solution

The only thermodynamic information derivable from a homogeneous solution containing $n_2$ moles of solute in $n_1$ moles of solvent is that, 1. the change in the Gibbs free energy of a solution is given by

$$dG = -SdT + Vdp + \frac{\partial G}{\partial n_1} dn_1 + \frac{\partial G}{\partial n_2} dn_2 \tag{6}$$

and 2. the change in the partial molar Gibbs free energy of the solvent is given by

$$d\mu_1 \equiv d\overline{G}_1 = -\overline{S}_1 dT + \overline{V}_1 dp + \frac{\partial \mu_1}{\partial x_2} dx_2 \tag{7}$$

where $S$ is the entropy of the solution, $\overline{S}_1$ is the partial molar entropy of the solvent, $V$ is the volume of the solution and $\overline{V}_1$ is the partial molar volume of the solvent. In both equations the change in pressure, $dp$, refers only to a change in an externally applied pressure. These equations may be integrated from $n_2 = 0$ to $n_2$ and $x_2 = 0$ to $\frac{n_2}{n_1 + n_2}$ in order to obtain the entire change in the Gibbs free energy, $\Delta G$, and the entire change in the chemical potential of the solvent, $\Delta \mu_1$, while forming the solution at constant temperature and pressure by adding $n_2$ moles of solute to $n_1$ moles of solvent. Thus,

$$\int d\mu_1 = \int_0^{x_2} \frac{\partial \mu_1}{\partial x_2} dx_2 \ . \tag{8}$$

This is an extraordinarily simple statement. However, it reveals nothing at all about the process by which the change in chemical potential occurs with the addition of solute to solvent. In the final Chapter we shall return to the interpretation of the change in chemical potential $\Delta \mu_1$.

# Chapter III   Toward Understanding the Colligative Properties of a Solution

An alteration in the solvent, attributable to the dispersal of finely divided substances throughout the solvent, induces the colligative properties of the solution. These properties of the solution include its osmotic pressure, its melting temperature and boiling temperature and the vapor pressure of the solvent. There is, however, no adequate theoretical treatment which provides a coherent account of these properties even for water solutions. We have reviewed in Chapter I the history of venerable thought concerning the colligative properties, especially the osmotic pressure of a solution; and, as we have noted, the foundations for understanding these properties have been frequently misunderstood. Therefore, we shall reexamine these foundations and derive what we believe to be the only admissible account of the colligative properties of a solution. We shall conclude that the alteration in the solvent, which is attributable to the dispersal of solute throughout the solvent, is an enhanced negative hydrostatic pressure in the solvent within the solution (35 b).

## I. Osmotic Pressure

We shall begin by considering the essential features of the solution to which we may attribute its osmotic pressure. A solution is contained in an osmometer as illustrated in Fig. 52. The osmometer consists of a cylinder with a semipermeable membrane at the bottom and filled with a solution. The cylinder is standing in a basin containing the solvent. At equilibrium the column of the solution reaches a height $h$ above the surface of the solvent. The solute particles or molecules are depicted by dots within the boundaries of the solution. The solvent in the solution and the liquid below the membrane is not depicted by any symbol, but its presence is implied. The solvent vapor is depicted by small dots above the pure solvent where its vapor pressure is a function of $z$, $p_V(z)$. No other gas is enclosed within the system. The temperature of the system remains constant at $TK$ and the system is at equilibrium. The solvent vapor and liquid and the solute molecules are subject to a gravity field with an acceleation of $g$. The mass and volume of an average solute particle are $m_2$ and $v_2$ respectively and the mass and volume of a solvent molecule are $m_1$ and $v_1$ respectively. The solvent has an average density $\rho_1 = m_1/v_1$. For illustration, the solvent may sometimes be designated as water. The solute is a solid non-volatile molecule or particle which does not dissociate into subunits in the solvent nor does it chemically react with the solvent. The density of the solute is $\rho_2 = m_2/v_2$ and is depicted in Fig. 52

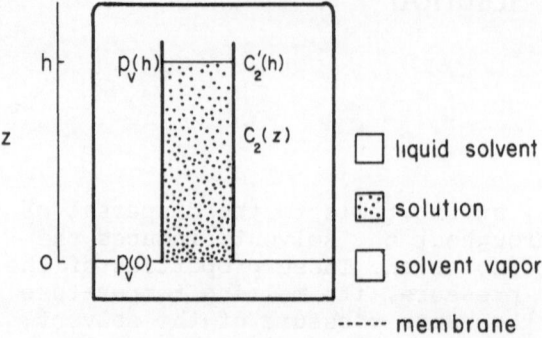

Fig. 52. Osmotic column at equilibrium in gravitational field. The solution is retained above a rigid, semipermeable membrane by an upright cylinder in a gravity field. The upper surface of the solution is exposed only to the vapor pressure of the solvent. The temperature is constant. At equilibrium the upper surface of the solution assumes a height $h$ above the surface of the pure solvent. At equilibrium and without convection the vapor pressure and the molar concentration of the solute in the solution are shown to be a function of $z$, $p_v(z)$ and $c_2(z)$, in a gravity field. $z$ is the level above pure solvent

to be considerably more than the average density of the solvent, i.e. $\rho_2 \gg \rho_1$. The membrane in Fig. 52 is shown to be impermeable to the solute, whereas it is permeable to the solvent. The membrane is also rigid so that it is not deformed by the pressure of the molecules it reflects.

Next we shall consider the properties of the solute, the pure solvent liquid, the solution and the solvent vapor in this system which enable us to predict the height of the column $h$ in relation to the molar concentration of the solute, $\bar{C}_2$, the number of moles of solute per liter of solution. The purpose of our presentation is to base our prediction on unquestioned physical principles applied to an idealized solution and thereby avoid the difficulties attending real solutes which may chemically react to some degree with the solvent or which may dissociate to some degree into sub units and avoid the uncertainty of real membranes which may be somewhat permeable to the solute.

To predict the precise equilibrium relationship between the molar concentration and the height of the solution in the osmometer in Fig. 52, there are several features of the system to be considered. *First*, in a gravity field, the vapor pressure above the pure solvent will diminish with increasing height above the level of the pure solvent at $z = 0$. *Second*, the molecules of the solute are in thermal motion so that they are dispersed throughout the solution and exert a pressure on all boundaries of the solution. The solvent molecules are also in thermal motion and will exert a pressure at all boundaries from which they are reflected. In a solution where the molecular spacing may be of the same order of magnitude as the molecular

diameter, the thermal motion of all molecules, solute and solvent, exerting a pressure at the boundary may be both translational and vibrational. *Third*, in a gravity field, the molar
concentration of the solute will be a function of $z$ and diminish with increasing height if, as depicted in Fig. 52, the density of the solute exceeds the density of the solvent. *Fourth*, the
solute molecules or particles at $h$ maintain a column of *solvent*
of unit area and a weight of $\rho_1 gh$. *Fifth*, the solvent is distensible[14] when subjected to tensile forces and tension in the
solvent lowers its vapor pressure. *Sixth*, the vapor pressure of
the solvent within the solution must be altered in some manner
by the solute molecules and to such an extent that, at $z = h$,
the vapor pressure of the solution exactly matches the vapor
pressure at height $h$ above the solvent in the basin. We shall
now consider each of these separate features and, in combination,
they will enable us to recognize the unique property of the solution to which we may attribute the osmotic pressure and the
other colligative properties of the solution. First we shall
consider how vapor pressure of the pure solvent varies with
$z$, $p_V(z)$.

## Distribution of Vapor Molecules in a Gravity Field

Assuming that the attractive or repulsive forces between molecules of vapor have a negligible effect on the pressure they
exert, the vapor pressure at any height $z$ induced by elastic
collisions between molecules will be

$$p_V(z)\left[Bdz - c_V(z)Bdzv_1\right] = c_V(z)BdzkT$$

where $Bdz$ is a volume at $z$ with cross section $B$, $c_V(z)$ is the
number of vapor molecules per unit volume at $z$, $k$ is the Boltzmann constant and $T$ is the absolute temperature. Although the
volume occupied by the vapor molecules, $c_V(z)Bdzv_1$, may be a
small portion of the volume $Bdz$, it may not be negligible. Therefore, the vapor pressure of such a vapor will be at $z$,

$$p_V(z) = \frac{c_V(z)kT}{1 - c_V(z)v_1} .$$

Rearranging this equation, the concentration of vapor at $z$ will
be

$$c_V(z) = \frac{p_V(z)}{kT + v_1 p_V(z)} .$$

Now the upward force per unit area at $z$, $p_V(z)$, is greater than
the downward force per unit area at $z + dz$, $p_V(z + dz)$, by the

---

[14]
  Distensibility is equivalent to negative compressibility

weight of the molecules in a volume of 1 unit area · $dz$. The number of vapor molecules in this volume is $c_v(z)dz$ and their weight is $c_v(z)dz\, m_1g$; therefore

$$dp_v(z) = -m_1g\, c_v(z)dz .$$

Eliminating $c_v(z)$ in this equation by using the previous equation,

$$\frac{kT + v_1 p_v(z)}{p_v(z)} dp_v(z) = -m_1gdz ;$$

and integrating between the limits $z = 0$ and $z$

$$\ln \frac{p_v(z)}{p_v(o)} = -\frac{v_1}{kT}\left[p_v(z) - p_v(o)\right] - \frac{m_1gz}{kT} . \qquad (9)$$

This equation is of course, an expression of the Boltzmann distribution law applied to gases in a gravity field. To insure that, in equilibrium, there is no perpetual distillation of water vapor to or from the solution surface at $h$, the solute must alter the vapor pressure of the solution at $h$ in Fig. 52 to precisely this value, i.e.

$$\ln \frac{p_v(h)}{p_v(o)} = -\frac{v_1}{kT}\left[p_v(h) - p_v(o)\right] - \frac{m_1gh}{kT} . \qquad (10)$$

## Thermal Pressure of the Solute and Solvent Molecules within the Solution

The second feature of the system depicted in Fig. 52 is that the solute particles or molecules must be dispersed (or dispersing) throughout all solvent from which they are not excluded. This dispersal is a consequence of the translational component of the thermal motion of all solute and solvent molecules. At equilibrium, as depicted in Fig. 52, the solute molecules are dispersed throughout all the solution standing in the column of cross section $A$ and height $h$ above the semipermeable membrane.

Wherever the path of a *solute* molecule intercepts a boundary of the solution, the molecule is reflected as momentum is transferred. A perpendicular force is thereby exerted on the boundary which is the sum of the perpendicular components of the momenta transferred by all such solute molecules per unit time. At all boundaries of the solution this force must be opposed. Similarly, the translational and vibrational components of the thermal motion of the *solvent* molecules exert a force wherever they are reflected at a boundary, and this force must also be opposed.

Binding forces between all the molecules of the solution balance the forces exerted by their combined thermal motion. Cohesive forces between the molecules of the solvent maintain it in the liquid phase. Likewise, there may be more or less equal adhesive forces between molecules of the solvent and molecules of the

solute. *The cohesive and adhesive forces between all molecules
and their thermal or Brownian motion are the essential ingredients on which to base a coherent account of all the colligative
properties of a solution.*

In order to ascertain the magnitude of the opposing tension in
the solvent and to predict precisely the height of the osmotic
column in Fig. 52 we shall discuss forces derived from the thermal motion of all molecules.

Before considering the magnitude of the thermal pressures of
the solute and solvent molecules, we need to define the concentration of the solute and solvent in Fig. 52. The solution was
constituted by adding $n_2$ moles of solute to $V_1$ liters of solvent
to make $V_S$ liters of solution at temperature $T$. The molar concentration of the solute in the solution is by definition
$\overline{C}_2 = \dfrac{n_2}{V_S}$ and the average tensile concentration of the solute is by
definition $\overline{C}_2' = \dfrac{n_2}{n_1 \overline{V}_1}$ (see glossary). At equilibrium the concentration of the solute becomes a function of $z$ in a gravity field
when $\rho_1 \neq \rho_2$. An incremental volume of solution, $dV_S = Adz$, is
the combined incremental volume of the solvent $dn_1(z)\overline{V}_1(z)$, and
incremental volume of solute, $dn_2(z)\overline{V}_2(z)$, at the height $z$, i.e.

$$Adz = dn_1(z)\overline{V}_1(z) + dn_2(z)\overline{V}_2(z) \tag{11}$$

so that

$$V_S = \int dV_S = \int_0^h Adz = \int dn_1(z)\overline{V}_1(z) + \int dn_2(z)\overline{V}_2(z) \ .$$

The concentration of the solute as a function of $z$ may be defined
in three ways: 1. as molar, 2. as tensile, 3. as molal:

1. The *molar concentration of solute* $c_2(z)$, is defined as the
number of moles of solute in a unit volume of solution at
height $z$, i.e.

$$c_2(z) = \frac{dn_2(z)}{Adz} = \frac{dN_2(z)}{N_A Adz} \tag{12}$$

where $dn_2(z)$ is the number of moles of solute in volume $Adz$ at
height $z$ and $dN_2(z)$ is the number of solute molecules in $Adz$
at $z$. Similarly, the number of moles of solvent in a unit volume
of solution at $z$ is the *molar concentration of solvent* $c_1(z)$,
i.e.

$$c_1(z) = \frac{dn_1(z)}{Adz} = \frac{dN_1(z)}{N_A Adz} \ .$$

Of course,

$$N_2 = \int_0^h c_2(z) N_A Adz \quad \text{and} \quad N_1 = \int_0^h c_1(z) N_A Adz$$

so that

$$\bar{C}_2 = \frac{\int_0^h c_2(z) A dz}{V_S} \quad \text{and} \quad \bar{C}_1 = \frac{\int_0^h c_1(z) A dz}{V_S} \quad . \tag{13}$$

Dividing Eq. (11) by $A dz$ and substituting for the definitions of $c_1(z)$ and $c_2(z)$ then

$$1 = c_1(z)\bar{V}_1 + c_2(z)\bar{V}_2 \tag{14}$$

and

$$dc_2(z) = \frac{\bar{V}_1}{\bar{V}_2} dc_1(z) \quad ,$$

assuming the dependence of $\bar{V}_1$ and $\bar{V}_2$ on $z$ is negligible. If the molar concentration of solute at $z$ decreases with increasing $z$, as in Fig. 52, then the molar concentration of solvent must increase by the ratio $\bar{V}_2/\bar{V}_1$ times the decrease in molar concentration of solute.

2. The *tensile concentration of the solute* $c_2'(z)$, is defined as the number of moles of solute in a unit volume of solvent at height $z$,

$$c_2'(z) = \frac{dn_2(z)}{dN_1(z)v_1} = \frac{dn_2(z)}{dn_1(z)\bar{V}_1(z)} \tag{15}$$

i.e. the tensile concentration of the solute in the solution at $z$ is the reciprocal of the partial molar volume of the solvent times the mole ratio of solute to solvent at $z$. Similarly, the number of moles of solvent in a unit volume of solvent at height $z$, the *tensile concentration of the solvent* $c_1'(z)$, is

$$c_1'(z) = \frac{dn_1(z)}{dn_1(z)\bar{V}_1(z)} = \frac{1}{\bar{V}_1(z)} \tag{16}$$

i.e. the tensile concentration of the solvent in the solution at $z$ is the reciprocal of the partial molar volume of the solvent at $z$. Combining Eq. (15) and (16)

$$c_2'(z) = \frac{dn_2(z)}{dn_1(z)} c_1'(z) \quad . \tag{17}$$

If the dependence of the partial molar volume of solvent on $z$ is negligible, i.e. $\bar{V}_1$ is constant, then $c_1'(z)$ is constant and the tensile concentration of solute varies with $z$ as the mole ratio of solute to solvent varies with $z$. The relationship between the tensile and molar concentrations of the solute at $z$ can be obtained from their respective definitions, Eq. (12) and (15)

$$c_2'(z)/c_2(z) = \frac{A dz}{dn_1(z)\bar{V}_1(z)} \quad .$$

From Eq. (11) this becomes

$$c_2'(z)/c_2(z) = 1 + \frac{dn_2(z)\overline{V}_2(z)}{dn_1(z)V_1(z)} = 1 + c_2'(z)\overline{V}_2(z) \tag{18}$$

or

$$c_2'(z) = \frac{c_2(z)}{1 - c_2(z)\overline{V}_2(z)} \ . \tag{19}$$

3. The *molal concentration of the solute* at $z$ is defined as

$$c_2''(z) = \frac{dn_2(z)}{dN_1(z)m_1} = \frac{dn_2(z)}{dn_1(z)M_1}$$

and the *molal concentration of the solvent* at $z$ is defined as

$$c_1''(z) = \frac{dn_1(z)}{dn_1(z)M_1} = \frac{1}{M_1} \ .$$

Again we note that

$$\frac{dn_2(z)}{dn_1(z)} = \frac{n_2(z)}{n_1(z)} = \frac{c_2''(z)}{c_1''(z)} = \frac{c_2'(z)}{c_1'(z)} = \frac{c_2(z)}{c_1(z)} \ .$$

At equilibrium in a gravity field, the molar concentration of both solute and solvent molecules will be a function of $z$ when $\rho_1 \neq \rho_2$. Therefore, it is imperative to determine the concentrations of solvent and solute at the unrestrained surface of the solution, i.e. $c_1(h)$ and $c_2(h)$ in Fig. 52. We shall determine these concentrations in a subsequent Section, but, before doing so, we must examine carefully our objective, which is to ascertain the magnitude of the induced tension in the solvent of the solution and compare it with the induced tension in the pure solvent.

## Induction of Enhanced Tension in the Solvent of a Solution

The crucial question is, how can the thermal motion of solute molecules enhance the solvent tension? Our first proposition is that van't Hoff's law can be extended to apply to all molecules in the solution and in the pure solvent and not restricted to a dilute solution. Van't Hoff (1887, 105) recognized that for dilute solutions a quantity he called the bombardment pressure is equal to $C_2RT$ where $C_2$ is the concentration of the solute, $R$ is the gas constant, and $T$ is the absolute temperature. He concluded that the osmotic pressure was equal to the bombardment pressure. We shall assert that the pressures exerted by the thermal motion of the solute and solvent molecules are proportional to their molar concentrations at the boundaries where they are reflected. We assert: 1. in the solution, the pressure $p_2(h)$ exerted by the *solute* molecules at the unrestrained surface of height $h$ is

$$p_2(h) = c_2(h)RT \ , \tag{20}$$

where $c_2(h)$ is the molar concentration of solute molecules at $h$, $R$ is the gas constant and equals $kN_A$ and $T$ is the absolute temperature; 2. in the solution, the pressure $p_1(h)$ exerted by the solvent molecules is

$$p_1(h) = c_1(h)RT ,\qquad (21)$$

where $c_1(h)$ is the molar concentration of solvent molecules at $h$; and 3. in the pure solvent the pressure $p_0(o)$ exerted at the surface of $z = 0$ is

$$p_0(o) = c_0(o)RT ,\qquad (22)$$

where $c_0(o)$ is the molar concentration of solvent molecules.

Our second proposition is that the total thermal pressure of all molecules at the surface at $h$ is the combined pressures $p_1(h) + p_2(h)$.

Our third proposition is that in the solution the cohesive force between all molecules must balance the thermal force of all molecules reflected from the surface at $h$. Likewise, in pure solvent the cohesive force must balance the thermal force of all molecules reflected from its surface at $z = 0$.

Our fourth and final proposition is that the tension induced in the solvent in either pure solvent or in a solution is the total force exerted by all molecules divided by the *area of the solvent* opposing these forces.

From these four propositions we shall deduce that 1. the solvent tension in the solution at $h$ is greater than in the pure solvent in the basin at $z = 0$, 2. the increased solvent tension at $h$ lowers the vapor pressure by the Poynting relation so as to match the Boltzmann distribution of the vapor emanating from the basin, and 3. no other property of the solvent at $h$ is altered (or could be altered) to account for the osmotic pressure and the lowered vapor pressure of the solution in the osmometer in Fig. 52.

In the *pure solvent* the thermal force exerted by reflection against a unit area of surface at 0 is $p_0(o)$. The opposing tension $\tau_0$ is this force divided by unit surface area, therefore

$$\tau_0(o) = p_0(o) = c_0(o)RT .\qquad (23)$$

This tension is transmitted throughout the solvent by cohesive force between all solvent molecules (Pascal's principle). The tension at a depth $-z$ below the surface is diminished by the hydrostatic pressure of the solvent between 0 and $-z$. Actually, the tension at $z = 0$ in Fig. 52 is slightly less than $p_0(o)$ by the externally applied vapor pressure of the solvent at $z = 0$. Likewise, the tension at a flat surface of pure solvent maintained at $z = h$ would also be slightly less than $p_0(o)$ by the vapor pressure at $h$[15]. Since the vapor pressures at o and $h$ produce a negligible change in the vapor pressure of pure solvent, we may state that

$$\tau_0(h) = \tau_0(0) = p_0(0) = p_0(h) = c_0(h)RT . \tag{24}$$

A unit area of the *solution* surface has fewer solvent molecules than pure solvent. The fraction of the unit area which is solvent in the solution is the fraction of the volume which is solvent, namely

$$\frac{n_1(h)\overline{V}_1}{n_1(h)\overline{V}_1 + n_2(h)\overline{V}_2} .$$

*The opposing tension $\tau_1(h)$ induced in the solvent must therefore be the combined outward forces exerted by solute and solvent per unit area of solution surface divided by the area of the solvent opposing these forces.*

The force exerted by solvent molecules against a unit area of solution is $p_1(h)$ and that exerted by solute molecules is $p_2(h)$; therefore

$$\tau_1(h) = \frac{[p_1(h) + p_2(h)]}{\left(\dfrac{n_1(h)\overline{V}_1}{n_1(h)\overline{V}_1 + n_2(h)\overline{V}_2}\right)} .$$

The tension in the solvent in the solution is greater than the tension in pure solvent by

$$\Delta\tau_1(h) = \tau_1(h) - \tau_0(h)$$

or

$$\Delta\tau_1(h) = [p_1(h) + p_2(h)]\frac{n_1(h)\overline{V}_1 + n_2(h)\overline{V}_2}{n_1(h)\overline{V}_1} - p_0(h) .$$

In our first proposition we assert that each of these pressures is equal to $c(h)RT$; therefore $p_1(h) = \dfrac{n_1(h)RT}{n_1(h)\overline{V}_1 + n_2(h)\overline{V}_2}$,

---

[15]At 25.0°C the vapor pressure of water at $z = 0$ would be 23.756 mmHg. According to the Boltzmann distribution, the vapor pressure at a height of 100 cm above the water in the basin would be 23.75431 mmHg. This small change in pressure, $\Delta p_\ell$, between 0 and 100 cm would also change the vapor pressure of pure water, $\Delta p_v$, according to the Poynting relation,

$$\Delta p_v = \overline{V}_\ell/\overline{V}_v \,\Delta p_\ell .$$

At 25°C, $\overline{V}_\ell/\overline{V}_v = 2.3\cdot10^{-5}$, that is, the effect of the Boltzmann distribution of vapor would be to diminish the vapor pressure of water by only 0.0023 percent of the change in vapor pressure from 0 to 100 cm in a gravity field, $g = 980$ cm/s$^2$. This effect is entirely negligible. The lower vapor pressure at height $h$ above the water in the basin accounts for the fact that any pure water supported at $h$ will eventually evaporate and then condense into the water in the basin.

$$p_2(h) = \frac{n_2(h)RT}{n_1(h)\overline{V}_1 + n_2(h)\overline{V}_2} \quad \text{and} \quad p_0(h) = \frac{n_0(h)RT}{n_0(h)\overline{V}_0} \quad . \quad \text{We conclude}$$

that the enhanced tension in the solution at $h$ is,

$$\Delta\tau_1(h) = \frac{RT}{\overline{V}_1}\frac{n_2(h)}{n_1(h)} + (\frac{RT}{\overline{V}_1} - \frac{RT}{\overline{V}_0}) \quad .$$

Therefore, for an ideal solution for which $\overline{V}_0 = \overline{V}_1$,

$$\Delta\tau_1(h) = \frac{RT}{\overline{V}_1}\frac{n_2(h)}{n_1(h)} \quad , \tag{25}$$

that is, the tension in the solvent in a solution at $h$ exceeds the tension in pure solvent by an amount equal to $\dfrac{RT}{\overline{V}_1}\dfrac{n_2(h)}{n_1(h)}$ .

This is an important conclusion and is the basis for a coherent account for all the colligative properties of a solution.

Since the definition of the tensile concentration of the solute at $z = h$ is

$$c_2'(h) = \frac{n_2(h)}{n_1(h)}\frac{1}{\overline{V}_1(h)} \quad ,$$

then the solvent tension in the solution exceeds that in pure solvent by

$$\Delta\tau_1(h) = c_2'(h)RT \quad . \tag{26}$$

If the solute molecules dissociate into ions or independent sub-units within the solvent, then $n_2(h)$ must refer to the number of moles of these subunits. Likewise, if solvent complexes with molecules of solute then $n_2(h)$ must refer to the number of moles of the complex units and $n_1(h)$ must refer to the number of moles of free solvent. For high solute concentrations, and especially for large solute molecules, the solute pressure may exceed $c_2(h)RT$ due to crowding so that these solute molecules exert a mechanical (or matrix) pressure as well as a thermal pressure. For these solutions,

$$\Delta\tau_1(h) > \frac{RT}{\overline{V}_1}\frac{n_2(h)}{n_1(h)} \quad .$$

In solutions for which the volume of the solution is less than the separate volumes of the solute and solvent, i.e. $V_s < V_1 + V_2$ , then $\overline{V}_1 < \overline{V}_0$ and

$$\Delta \tau_1(h) \; < \; \frac{RT}{\overline{V}_1} \frac{n_2(h)}{n_1(h)} \; .$$

However, for many solutions of even moderate concentration,

$$\Delta \tau_1(h) \; = \; \frac{RT}{\overline{V}_1} \frac{n_2(h)}{n_1(h)} \; .$$

The enhanced tension in the solvent of a homogeneous ideal solution can be written as follows:

$$\Delta \tau_1(V_s - n_2 \overline{V}_2) \; = \; n_2 RT \; .$$

This is in the form of an equation of state of $n_2$ moles of solute molecules occupying a volume $n_2 \overline{V}_2$ in the solution whose volume is $V_s$. $\Delta \tau_1$, the additional tension in the solvent induced by and opposing the thermal motion of the solute molecules, is analogous to the wall pressure exerted by the container of $n_g$ moles of real gas occupying a volume $n_g \overline{V}_g$ in a container of volume $V$. The wall pressure induced by and opposing the thermal motion of the real gas, $p_w$, is

$$p_w(V - n_g \overline{V}_g) \; = \; n_g RT \; .$$

This is, of course, the equation of state of a real gas for which the pressure exerted by the gas is proportional to its molar concentration, $p_g = C_g RT = \dfrac{n_g RT}{V}$. We note that the opposing pressure exerted by the containing wall is greater than $p_g$ by the ratio $\dfrac{V}{V - n_g \overline{V}_g}$ .

To predict the value of $c_2'(h)$ and therefore $\Delta \tau_1$, we need to consider the third feature of the system, namely, the distribution of solute molecules in a gravity field.

## Distribution of Solute Molecules within a Solution in a Gravity Field

According to Archimedes' principle, the gravitational force, $m_2 g$, acting upon a particle of mass $m_2$ immersed in a liquid is diminished by the weight of the liquid it displaces. When the volume of the particle is $v_2$, the weight of the liquid it displaces is $v_2 \frac{m_1}{v_1} g$ where $m_1$ is the mass of the liquid molecule and $v_1$ is the volume which it occupies. The net downward force, after the buoyant force has been subtracted from the gravitational force, is:

$$m_2' g \equiv m_2 g - \frac{v_2}{v_1} m_1 g = m_2 \left(1 - \frac{v_2 m_1}{m_2 v_1}\right) g \ . \tag{27}$$

In the solution in Fig. 52, $\frac{m_2}{v_2} \gg \frac{m_1}{v_1}$ so that the solute molecules will be distributed in accordance with Boltzmann's law, as are any molecules or particles dispersed by their thermal motion, e.g. the vapor molecules already discussed. The vertical force exerted by the solute molecules at $z$ from below is $Ap_2(z)$ and the vertical force from above at $z + dz$ is a lesser force and is $Ap_2(z + dz)$. The difference in these forces, $Adp_2(z)$, must be due to the net weight of the solute molecules in the volume $Adz$. The net weight of each particle or molecule is $m_2' g$ and there are $dN_2(z)$ molecules in the volume $Adz$, so that,

$$Adp_2(z) = -m_2' g c_2(z) N_A A dz \ .$$

Now the change in pressure is also

$$dp_2(z) = p_2(z + dz) - p_2(z) = RT \left[ c_2(z + dz) - c_2(z) \right] = RT dc_2(z) \ .$$

So we may combine the last two equations; thus

$$dc_2(z) = -\frac{m_2' g}{kT} c_2(z) dz \ .$$

Integrating this equation,

$$c_2(z) = c_2(o) e^{-\frac{m_2' g z}{kT}} \ . \tag{28}$$

Recalling the definition of the molar concentration of the solute in the solution in Fig. 52, Eq. (12), then

$$\int_o^h c_2(z) dz = \frac{V_S}{A} \overline{C}_2 = h \overline{C}_2 \ ;$$

so that

$$h \overline{C}_2 = -\frac{kT}{m_2' g} c_2(o) \left[ e^{-\frac{m_2' g h}{kT}} - 1 \right] \ .$$

Solving this equation for $c_2(o)$ and substituting into Eq. (28) the molar concentration of the solute at $z = h$ can be determined from

$$c_2(h) = \frac{\frac{m_2'gh}{kT}\, \bar{C}_2\, e^{-\frac{m_2'gh}{kT}}}{1 - e^{-\frac{m_2'gh}{kT}}} \cdot \tag{29}$$

## The Weight of the Column Below the Surface of the Solution and its Correlation with the Concentration at the Free Surface and the Osmotic Pressure of the Solution

Noyes (1900) proposed an experiment which illustrates that the solute alters the solvent in a solution in such a way so as to support the entire weight of *the solvent* below the surface of the solution in Fig. 53 (reference Fig. 33 for original Noyes column). Our version of the Noyes experiment compares the condition of the solvent in the solution in the left column in Fig. 53 with the condition of the solvent in the solution in the right column. The rigid semipermeable membrane for the osmometer on the right is only a distance $\Delta h$ below the surface of the solution at $h$. We require that the molar concentration of the solute at $h$, $c_2(h)$, be exactly the same as the molar concentration at $h$ in the left column in Fig. 53, so that the vapor pressures of the solvent in both columns are the same. We then let $\Delta h$ approach zero so that the hydrostatic pressure in the water below the membrane is $\bar{\rho}_w gh$ less than at $z = 0$ where $\bar{\rho}_w$ is the average density of the water between $z = 0$ and $h$. $\bar{\rho}_w gh$ is, of course, the weight of a column of water of unit cross section in a gravity field and it must equal in magnitude the osmotic pressure, $\Pi$, of the solution above the membrane at $z = h$. $\Pi$ also equals the osmotic pressure of the solution in the left column. This version of the Noyes experiment illustrates two important points. 1. Regardless of the distribution of solute molecules within the solution in a gravity field, at equilibrium the osmotic pressure of the solution is determined by the solute concentration at the unrestrained surface of the solution. 2. When the osmotic pressure of the solution achieves equilibrium with pure solvent in a gravity field, as illustrated in Fig. 53, the osmotic pressure equals the weight of a unit cross section of the *solvent* between the free surface of the solution and the free surface of the pure solvent, i.e. $\Pi$ is always exactly equal $\bar{\rho}_w gh$. This remains true whether the solvent in the right column (or in the left column) in Fig. 53 is distensible or not. As we shall next demonstrate, even when the solvent is distensible, it is the exact equality between $\Pi$ and $\bar{\rho}_w gh$ that insures that the vapor pressure above the solution at $h$ matches the vapor pressure a distance $h$ above the pure solvent.

## Vapor Pressure of a Distensible Liquid under Tension

The relationship between the vapor pressure of a liquid and the hydrostatic pressure to which the liquid is subjected is given by a thermodynamic statement attributed to Poynting, namely,

$$\int \bar{V}_\ell dp_\ell = \int \bar{V}_v dp_v \tag{1}$$

where $\overline{V}_\ell$ is the molar volume of the liquid, $p_\ell$ is the pressure (or tension) within the liquid, $\overline{V}_v$ is the molar volume of the gaseous vapor and $p_v$ is the vapor pressure. (See p. 43, 77 for a derivation of the Poynting relation). This statement is the same as stating that at constant temperature, changing the pressure on a liquid from $p_\ell$ to $p_\ell + \Delta p_\ell$ must alter the vapor pressure with which the liquid is in equilibrium from $p_v$ to $p_v + \Delta p_v$ such that the change in the Gibbs molar free energy is the same in both the liquid and its vapor. If the temperature remains constant, then the change in the Gibbs molar free energy (also designated as the change in chemical potential) is due only to the pressure-volume work in the liquid and in the vapor.

The hydrostatic pressure in the solvent (water) in either the left or the right column in Fig. 53 varies from $p_\ell(o) = p_v(o)$ at $z = o$ to $p_\ell(h) = p_v(o) - \overline{\rho}_w gh$ at $z = h$, so that

$$p_\ell(h) - p_\ell(o) = -\overline{\rho}_w gh .\tag{30}$$

With the Poynting relation, we can determine how the vapor pressure of the solvent in the right column must change from $z = o$ to $h$ and how the vapor pressure of the solvent in the solution in the left column must change from $z = o$ to $h$. If the solvent is distensible when subjected to tension, then the partial molar volume, $\overline{V}_\ell(z)$, will vary from $z = o$ to $h$. However, we need to evaluate the left integral of the Poynting relation between the limits $p_\ell(o)$ and $p_\ell(h)$. Even though $\overline{V}_\ell(z)$ may vary somewhat with $z$, the left integral, integrated between its limits, is simply the average molar volume of the liquid solvent times the difference in pressures at the limits, i.e.

$$\int_{p_\ell(o)}^{p_\ell(h)} \overline{V}_\ell(z) \, dp_\ell(z) = \overline{\overline{V}}_\ell \left[ p_\ell(h) - p_\ell(o) \right]\tag{31}$$

or combining Eqs. (30) and (31)

$$\int_{p_\ell(o)}^{p_\ell(h)} \overline{V}_\ell(z) \, dp_\ell(z) = -\overline{\overline{V}}_\ell \overline{\rho}_\ell gh .$$

Now the product of the molar volume and the density at any height $z$ is always, by definition,

$$\overline{V}_\ell(z) \rho_\ell(z) = \frac{N_1(z) v_1}{n_1(z)} \cdot \frac{n_1(z) N_A m_1}{N_1(z) v_1} = N_A m_1 .$$

Therefore, the product of average molar volume and average density of the solvent must be $N_A m_1$ and the left integral of the Poynting equation becomes

$$\int_{p_\ell(0)}^{p_\ell(h)} \overline{V}_\ell(z) \, dp_\ell(z) = -m_1 g N_A h .$$

(32)

To evaluate the right integral of the Poynting equation, we refer to the semi-ideal law which includes the fact that a molecule of vapor has a finite volume $v_1$ and assumes only that the forces between molecules of vapor have a negligible effect upon their thermal pressure; thus

$$p_V(z) \left[ B dz - dn_V(z) N_A v_1 \right] = dn_V(z) N_A kT$$

(33)

where $B$ is the cross sectional area of a column of vapor and $dn_V(z)$ is the number of moles of solvent vapor in volume $B dz$. The molar volume of the vapor at $z$ is

$$\overline{V}_V(z) = \frac{B dz}{dn_V(z)}$$

so that $B dz$ in Eq. (33) may be replaced and $p_V(z)$ becomes

$$p_V(z) = \frac{N_A kT}{\overline{V}_V(z) - N_A v_1}$$

or $\overline{V}_V(z)$ is

$$\overline{V}_V(z) = N_A v_1 + \frac{N_A kT}{p_V(z)}$$

(34)

The right integral of the Poynting relation becomes, with Eq. (34)

$$\int_{p_V(0)}^{p_V(h)} \overline{V}_V dp_V(z) = N_A v_1 \left[ p_V(h) - p_V(0) \right] + N_A kT \, \ell n \frac{p_V(h)}{p_V(0)} .$$

(35)

Combining Eqs. (32) and (35), we obtain from the Poynting relation that

$$\ell n \frac{p_V(h)}{p_V(0)} = - \frac{v_1}{kT} \left[ p_V(h) - p_V(0) \right] - \frac{m_1 g h}{kT} .$$

This equation is precisely the same as Eq. (10) for the vapor pressure at a height $h$ above the pure solvent which was derived as the Boltzmann distribution law. If the volume occupied by the vapor molecules is negligible, this equation simplifies to

$$\ell n \; \frac{p_v(h)}{p_v(o)} = - \; \frac{m_1 g h}{kT}$$

or since $m_1/k = \overline{\overline{V_1}} \bar{\rho}_1/R$

$$- \; \frac{RT}{\overline{\overline{V_1}}} \; \ell n \; \frac{p_v(h)}{p_v(o)} = \bar{\rho}_1 g h \tag{36}$$

## Summary of the Conditions which Describe the Solution in a Gravity Osmometer shown in Figs. 52 and 53

We have not yet directly stated what property of the solvent has been altered by the thermal motion of the solute to which we must attribute the colligative properties of the solution above the semipermeable membrane in Figs. 52 and 53. We now know, however, the essential facts which accompany the altered state of the solvent. The solvent in the solution in Figs. 52 and 53 is altered such that, at equilibrium, 1. the concentration of solute molecules at the upper unrestrained surface of the solution determines the height of the solution above the semipermeable membrane, regardless of the distribution of solute molecules *elsewhere* in the solution, 2. the osmotic pressure of the solution is determined by the concentration of solute molecules at the free surface, regardless of the concentration elsewhere, 3. the osmotic pressure equals the weight of a unit column of *solvent* between the free surfaces of the solution and the pure solvent, i.e. $\bar{\rho}_1 g h$, 4. for the ideal solution for which the volume of the solution is the exact sum of the separate volumes of

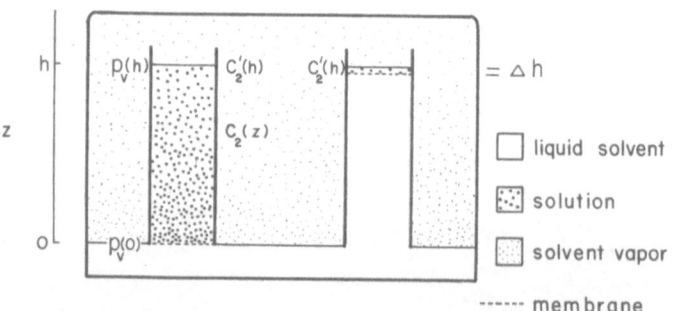

Fig. 53. Role of solute concentration at the free surface. Osmometer $A$ is the same as in Fig. 52. In $B$ the semipermeable membrane has been elevated to a distance $\Delta h$ below the equilibrium height $h$ in $A$. The osmotic concentration $c_2'(h)$ is the same in both osmometers. Solute, solvent and vapor are in gravitational equilibrium $\tau_\ell$. At the freezing point the vapor pressure from the tensile water matches that of the ice

the solute and the solvent, the concentration at the free surface is determinable provided the relative densities of solute
and solvent are known, and 5. the combined dispersal forces of
solute and solvent molecules are additive and are balanced by
the cohesion of fewer ( $\dfrac{n_1(h)\overline{V}_1}{n_1(h)\overline{V}_1 + n_2(h)\overline{V}_2}$ ) solvent molecules in
the surface at $h$, i.e. the solvent tension in the ideal solution
exceeds that in the pure solvent by $\Delta\tau_1 = \dfrac{RT}{V_1}\dfrac{n_2(h)}{n_1(h)} = c_2'(h)RT$.

When describing the altered state of the solvent in the solution
in accordance with the first three conditions, we insure that
this satisfies precisely the Boltzmann distribution law and the
Poynting relation for solutions whose solute molecules differ in
density from the solvent.

## The Greater Tension in the Solvent within a Solution Alters its Osmotic Pressure with Respect to Pure Solvent

We now inquire into the property of the solvent which is altered
by the solute in such a way and to such an extent as to sustain
the free surface of the solution at $z = h$ and to lower the vapor
pressure of the solvent within the solution at $h$ to precisely
the value it must be in order to match the vapor pressure of
the solvent at the height $h$ above the pure solvent. Hulett
(1902, 41) was the first to clearly recognize the manner by
which the solute alters the solvent. He recognized that the
weight of the solvent in the solution at $h$ must oppose the thermal pressure exerted by the solute molecules at $h$. By opposing
this solute pressure a negative pressure is induced in the solvent at $h$. Hulett also recognized that it was the tension or
negative hydrostatic pressure within the solvent which lowered
its vapor pressure to the precise value required by the Poynting
relation and thereby the Second Law of Thermodynamics, namely
the vapor pressure over the solution and over the pure solvent
at $z = h$ must match exactly. Even when the solute molecules dissociate or hydrate or even when crowding or other mechanical
forces exert a pressure which is added to the thermal pressure
of the solute molecules, so that the solvent tension at the surface exceeds the value determined by $c_2'(h)RT$, i.e.

$$\Delta\tau_1 > c_2'(h)RT,$$

nevertheless, the enhanced tension in the solvent equals the osmotic pressure and equals the weight of a unit column of solvent
of height $h$, i.e.

$$\Delta\tau_1 = \Pi = \overline{\rho}_1 gh . \tag{37}$$

We should note that the osmotic pressure, which equals the enhanced tension in the solvent, is more closely approximated to
the molal concentration than it is to the molar concentration
in an aqueous solution. We have assumed that the thermal pressure of the solute molecules is proportional to their molar concentration (cf. Eq. (20)).

## Height of Solution Column in a Cylindrical Gravity Osmometer

Finally, we may predict the height of the column of the solution in Figs. 52 and 53 knowing only the average molar concentration $\bar{C}_2$. Combining Eqs. (26) and (37)

$$\bar{\rho}_1 gh = c_2'(h) RT \ . \tag{38}$$

Combining Eqs. (19), (27) and (29) to obtain a value for $c_2'(h)$ and then substituting this value into Eq. (38) and simplifying the resultant equation, we obtain

$$e^{\dfrac{M_2(\rho_2-\rho_1)gh}{\rho_2 RT}} = 1 + \frac{M_2(\rho_2-\rho_1)\bar{C}_2}{\rho_1\rho_2} + \frac{M_2(\rho_2-\rho_1)\bar{V}_2\bar{C}_2 gh}{\rho_2 RT} \tag{39}$$

With this equation and knowing the molar concentration of the solution, the densities of solute and solvent and the molecular weight and molar volume of the solute, a precise value of the equilibrium height $h$ in Figs. 52 and 53 may be determined. This value assumes that the solute and solvent obey Eq. (14) i.e. form an ideal solution, and that the solute does not dissociate in the solvent nor complex with the solvent.

Suppose a solution of inulin containing 50 g of inulin per 1000 g solution has a density $D_4^{20} = 1.0167$ g/cm$^3$ and the average molecular weight of the inulin is $M_2 = 5200$ g/mol (data from CRC Handbook of Chemistry and Physics). Assuming these data are accurate and assuming no hydration then for this solution at $0°C$,

$$\bar{C}_2 = \frac{50/5200}{1.000/1.0167} = 0.009776 \text{ mol/liter solution} \ .$$

The volume of 1000 g of solution $= \dfrac{1000}{1.0167} = 983.57$ cm$^3$. Since there are 950 g of water in the solution and its volume is 950 cm$^3$, then assuming the solution is ideal, the volume of 50 g of inulin is 33.57 cm$^3$. Since 50 g of inulin is $\dfrac{50}{5200}$ mol, the molar volume of inulin is $\bar{V}_2 = 33.57/.00962 = 3492$ cm$^3$/mol; so that $\bar{V}_2\bar{C}_2 = 0.03414$. The density of the inulin is $\rho_2 = \dfrac{5200}{\bar{V}_2} = 1.489$ g/cm$^3$. According to Eq. (39), the equilibrium height of the column above the membrane in Fig. 53 for this solution at $0°C$ is

$$h = 232.5 \text{ cm}$$

in a field of 1 g acceleration. In this example, the ratio $\dfrac{M_2(\rho_2-\rho_1)gh}{\rho_2 RT}$ equals 0.01753. The exponential term on the left side of Eq. (39) may be expanded into the series

$$1 + \frac{M_2(\rho_2-\rho_1)gh}{\rho_2 RT} + \frac{1}{2!}\left[\frac{M_2(\rho_2-\rho_1)gh}{\rho_2 RT}\right]^2 + \frac{1}{3!}\left[\frac{M_2(\rho_2-\rho_1)gh}{\rho_2 RT}\right]^3 + \cdots$$

Neglecting second order and higher terms, Eq. (39) reduces to an approximate relationship

$$\rho_1 gh \approx \frac{\bar{C}_2}{1 - \bar{V}_2 \bar{C}_2} RT \ ,$$

where $\frac{\bar{C}_2}{1 - \bar{V}_2 \bar{C}_2} = \bar{C}_2'$, according to Eq. (19). Thus, for the inulin solution in our example

$$h \approx 234.6 \text{ cm} \ .$$

The latter value is, of course, the height for a 0.009776 molar ideal solution for which $\rho_2 = \rho_1$ since for such a solution

$$h = \frac{\bar{C}_2' RT}{g \rho_1} \ ,$$

i.e. Eq. (38) for which $c_2'(h) \equiv \bar{C}_2'$ .

## II. Vapor Pressure

The vapor pressure of the solvent in a solution is lowered for the same reason that a solution exhibits an osmotic pressure. The enhanced hydrostatic tension in the solvent, induced by the thermal pressure which the solute molecules exert against the free boundary of the solution, lowers the vapor pressure of the solvent in the solution. When considering the lowering of the vapor pressure at h immediately above the surfaces of the two columns in Fig. 53, we recall that Hulett recognized that the only way to insure a match between the vapor pressure of the solvent above the solution at $z = h$ and the vapor pressure above the pure solvent at $z = h$ was to recognize that a tension was induced in the solvent at $z = h$ equal to $\bar{\rho}_1 gh$ and equal to $\Pi$ and, as we state, equal to $c_2'(h)RT$. From the Poynting relation and the combination of Eqs. (36) and (37) we obtained the familiar relationship for the vapor pressure,

$$\Pi = \bar{\rho}_\ell gh = -\frac{RT}{\bar{V}_1} \ln \frac{p_V(h)}{p_V(o)} \ ,$$

assuming we may neglect the volume of the vapor molecules.

If the solution is ideal so that the osmotic pressure is

$$\Pi = c_2'(h)RT$$

and if the density of the solute does not differ from the density of the solvent so that

$$c_2'(h) = \bar{C}_2' = \frac{n_2}{n_1 \bar{V}_1}$$

then

$$\ell n \ \frac{p_V(h)}{p_V(o)} = -\frac{n_2}{n_1} \ . \tag{40}$$

The ratio of vapor pressures of solution to pure solvent may be written

$$\frac{p_V(h)}{p_V(o)} = 1 + \frac{\Delta p_V}{p_V} \ .$$

If the change in vapor pressure of the solvent with the addition of solute is small, as it will be for dilute solutions, then the logarithmic term in Eq. (40) may be expanded in a series. Neglecting higher terms, we obtain the familiar approximation that

$$\frac{\Delta p_V}{p_V} \approx -\frac{n_2}{n_1}$$

or the change in vapor pressure of the solvent is approximately the mole ratio times the vapor pressure of the pure solvent for dilute solutions.


## III. Melting Point


In the introduction to Chapter II, we implied that all colligative properties of a solution are to be attributed to the negative hydrostatic pressure induced in the solvent by the thermal pressure exerted by solute and solvent molecules at the free surface. The way the melting point of a pure crystal of ice, immersed in the solution, is altered by the solute is illustrated in Fig. 54. Ice is shown as a crystal inclusion within the solution in cylinder B, and is also indicated exposed to its vapor in cylinder C. Pure water is shown below the membrane in B, and below the porous separator in A. Water and solute molecules are shown to co-exist uniformly within the boundaries of the solution and beyond the surfaces of the ice. Finally, water vapor is shown above the pure water on the left, above the solution in the middle and above the ice on the right. The temperature of all cylinders is the same and has been adjusted to $T_{\Pi_m}$, the melting temperature of the ice in the solution. To achieve equilibrium and equality of vapor pressures throughout, the pressures applied to the frictionless pistons at $g = o$ must be as follows. A positive pressure $p_{osv}$, equal to the vapor pressure of the ice at $T_{\Pi_m}$ is applied to the piston on the right and the pressure applied to the middle and left pistons must be changed from the positive vapor pressure of the water at $T_{\Pi_m}$ to a negative pressure, $-p_\ell$, such that

$$-\Pi = -p_\ell - p_{o\ell v} \ ,$$

where $\Pi$ is the osmotic pressure of the solution and $p_{o\ell v}$ is the vapor pressure of water subject only to its own vapor pressure at $T_{\Pi_m}$. Of course, the water must be free of nucleation centers so that it may remain at metastable equilibrium with respect to the negative hydrostatic pressure $-p_\ell$ and the temperature $T_{\Pi_m}$ which is below the freezing point $T_m$, of pure water, a condition that occurs frequently in the xylem sap of evergreen plants in winter (34, 35a).

To determine the melting point of an ice crystal in the solution, $T_{\Pi_m}$, that is, the temperature at which the vapor pressure of the ice equals the vapor pressure of the water in the solution, we must know how the vapor pressures of water and ice depend on temperature and pressure, and we must know the pressures applied to the water and the ice within the solution. The relationships are illustrated in the graphs on the right of Fig. 54. However, we shall derive them quantitatively.

The vapor pressure, $p_v$, of a liquid depends upon its temperature according to the thermodynamic statement, the Clapeyron equation derived on p. 79, Eq. (3)

$$\frac{dp_v}{dT} = \frac{\Delta \overline{H}_{\ell v}}{T(\overline{V}_v - \overline{V}_\ell)} \, , \tag{3}$$

where $\Delta \overline{H}_{\ell v}$ is the change in the molar enthalpy in transition from the liquid to vapor phase, $\overline{V}_v$ and $\overline{V}_\ell$ are the molar volumes of the vapor and liquid respectively and under the condition that the change in the external pressure applied to the liquid is the change in its vapor pressure.

If the liquid and its vapor are such that the molar enthalpy is constant, or nearly so, over a short range of temperatures, and if the vapor has the equation of state of an ideal gas, $\overline{V}_v = \frac{RT}{p_v}$, then neglecting the very small volume $\overline{V}_\ell$ compared with $\overline{V}_v$, the approximate relation known as the Clausius–Clapeyron equation is obtained,

$$\frac{d \ell n p_v}{dT} = \frac{\Delta \overline{H}_{\ell v}}{RT^2} \, . \tag{41}$$

Upon integration this becomes

$$p_v = C_{\ell v} \, e^{-\dfrac{\Delta \overline{H}_{\ell v}}{RT}} \, , \tag{42}$$

where $C_{\ell v}$ is the constant of integration. A similar expression can be derived for the vapor pressure of the solid solvent as a function of temperature,

$$p_v = C_{sv} e^{-\frac{\overline{H}_{sv}}{RT}} . \tag{43}$$

At the melting temperature of the pure solid solvent, $T_m$, these two vapor pressures are equal. If the equilibrium pressure of the vapor at the melting temperature has been determined experimentally, when the solid and liquid are subject only to the pressure of vapor $p_{ovm}$, then the constants $C_{\ell v}$ and $C_{sv}$ can be determined. Thus, the equations for the vapor pressure of pure water and ice as a function of temperature, when the ice and water are subject to the pressure of their vapor only, become respectively,

$$p_{o\ell v} = p_{ovm} e^{\frac{\Delta\overline{H}_{\ell v}}{R} \left[\frac{1}{T_m} - \frac{1}{T}\right]} \tag{44}$$

and

$$p_{osv} = p_{ovm} e^{\frac{\Delta\overline{H}_{sv}}{R} \left[\frac{1}{T_m} - \frac{1}{T}\right]} \tag{45}$$

Both these equations are depicted in graphic form in Fig. 54.

The dependence of the vapor pressure of the water as a function of the pressure applied to the water is given by the Poynting relation,

$$\int_{p_{o\ell v}}^{p_\ell} \overline{V}_\ell dp_\ell = \int_{p_{o\ell v}}^{p_{\ell v}} \overline{V}_v dp_v .$$

Assuming that water is incompressible and that water vapor is an ideal gas, then this relationship becomes

$$\overline{V}_\ell (p_\ell - p_{o\ell v}) = RT \ln\frac{p_{\ell v}}{p_{o\ell v}} ;$$

or the vapor pressure of water changes from $p_{o\ell v}$ to $p_{\ell v}$ when the pressure on the water changes from $p_{o\ell v}$ to $p_\ell$ according to

$$p_{\ell v} = p_{o\ell v}\, e^{\dfrac{\overline{V}_\ell}{RT}(p_\ell - p_{o\ell v})} \quad . \tag{46}$$

Similarly the vapor pressure of ice changes from $p_{osv}$ to $p_{sv}$ when the pressure on the ice changes from $p_{osv}$ to $p_s$ according to

$$p_{sv} = p_{osv}\, e^{\dfrac{\overline{V}_s}{RT}(p_s - p_{osv})} \quad . \tag{47}$$

Eqs. (44), (45), (46) and (47) can now be used to determine how the melting point varies with pressure applied to water and ice.

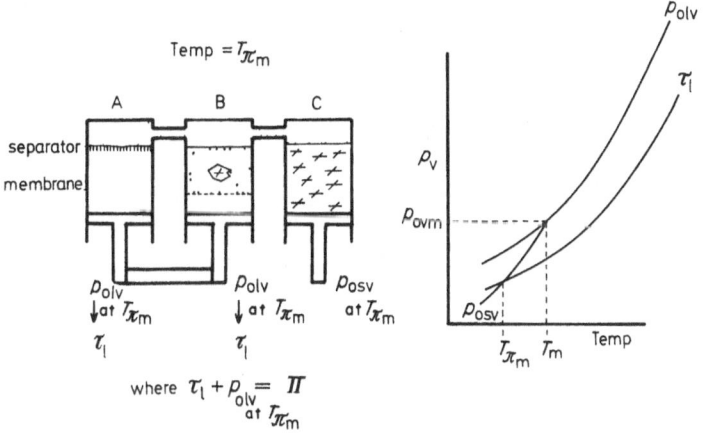

Fig. 54. Melting temperature lowered by solvent tension. $A$, cylinder containing water separated from its vapor by rigid porous membrane. $B$, solution with free surface and separated from solvent by semipermeable membrane. $C$, cylinder with fixed ice surface. $A$, $B$, $C$ are connected and have the same vapor pressure. Solvent in $A$ and $B$ are under tension

To determine the melting point, $T_{\Pi m}$, of a crystal of ice in the solution in Fig. 54 we also may apply these four equations. $T_{\Pi m}$ is that temperature at which $p_{\ell v}$ equals $p_{sv}$. Thus, by combining Eqs. (44) and (45) with Eqs. (46) and (47), at $T_{\Pi m}$

$$p_{ovm}\, e^{\dfrac{\Delta\overline{H}_{\ell v}}{R}\left[\dfrac{1}{T_m}-\dfrac{1}{T_{\Pi m}}\right]}\, e^{\dfrac{\overline{V}_\ell}{RT_{\Pi m}}(p_\ell - p_{o\ell v})} = p_{ovm}\, e^{\dfrac{\Delta\overline{H}_{sv}}{R}\left[\dfrac{1}{T_m}-\dfrac{1}{T_{\Pi m}}\right]}\, e^{\dfrac{\overline{V}_s}{RT_{\Pi m}}(p_s - p_{osv})} \quad .$$

To determine the melting temperature of the solution, we need to know the change in pressure induced in the water by the solute, $(p_\ell - p_{o\ell v})$, and the change in pressure applied to the ice by the solution $(p_s - p_{osv})$. As we now recognize, the thermal pressure of the solute and water molecules, induced an opposing negative pressure, $-p_\ell$, in the water within the solution and the change in the solvent pressure from $p_{ov\ell}$ to $-p_\ell$ is $(-p_\ell - p_{ov\ell}) = -\Pi$, the negative of the osmotic pressure. The ice crystal is an inclusion within the solution; and at its boundary with the solution, the solute and water molecules are reflected and exert a positive pressure which in an ideal solution equals that exerted by pure water at the crystal surface. Therefore the pressure on the crystal is not altered from $p_{osv}$ and $(p_s - p_{osv}) = 0$. We may remark here that no pressure is exerted on the wall of the cylinder by the thermal motion of the solute molecules for the same reason. Similarly, a bubble inclusion within the solution is not caused to expand by the tension induced in the water. In fact, if the radius of the inclusion is small, the bubble or the ice inclusions may be subject to a pressure $2\sigma_{\ell v}/r$ or $2\sigma_{\ell s}/r$, where $\sigma_{\ell v}$ and $\sigma_{\ell s}$ are the surface tension of the water with respect to vapor and to ice, respectively, and $r$ is the radius of the inclusion. The last equation can now be written

$$(\Delta \bar{H}_{sv} - \Delta \bar{H}_{\ell v}) \left[\frac{1}{T_m} - \frac{1}{T_{\Pi_m}}\right] = -\frac{\Pi \bar{V}_\ell}{T_{\Pi_m}} . \tag{48}$$

Since

$$(\Delta \bar{H}_{sv} - \Delta \bar{H}_{\ell v}) = \Delta \bar{H}_{s\ell} ,$$

the partial molar enthalpy for melting (the heat of fusion) and since the osmotic pressure for an ideal solution is

$$-\Pi = -\frac{RT}{\bar{V}_\ell} \frac{n_2}{n_1} ,$$

then we may write

$$T_{\Pi_m} = \frac{T_m}{1 + \frac{RT_m}{\Delta \bar{H}_{\ell s}} \frac{n_2}{n_1}} . \tag{49}$$

For 1 mole of ideal solute in 55.5009 moles of $H_2O$, $\frac{n_2}{n_1} = \frac{1}{55.5009}$, $R = 8.3143$ J/K·mol. For water $\Delta \bar{H}_{s\ell} = 5.983 \cdot 10^3$ J/mol and $T_m = 273.1598$ K at 1 atm pressure or $T_m = 273.1696$ at vapor pressure. Therefore

$$T_{\Pi_m} = 271.3140 \text{ K at vapor pressure}$$

or

$$T_{\Pi_m} = 271.3042 \text{ K at 1 atm pressure}$$

and the lowering of the melting temperature is

$$\Delta T_{\Pi_m} \equiv T_{\Pi_m} - T_m = -1.8556 \text{ K}.$$

The melting point is lowered by $1.8556°C$ by one mole of solute in 1000 ml of water; and the lowering of the melting temperature, like other colligative properties, is attributed to the enhanced tension in the water induced by the thermal pressure of the solute molecules exerted against the free surface of the solution.


## IV. Boiling Point

The boiling temperature of a solution, $T_{\Pi_b}$, is the temperature at which the vapor pressure of its solvent becomes equal to 760 mmHg (the pressure of one atmosphere or $1.01325 \cdot 10^6$ dyn cm$^{-2}$). Since the effect of the solute is to induce a tension in the solvent which reduces its vapor pressure at any and all temperatures, the solution will not boil at the boiling temperature, $T_b$, of its pure solvent.

The vapor pressure of the pure solvent subject to its own vapor pressure depends upon temperature in accordance with the Clapeyron equation. If the change in molar enthalpy for the transition from the liquid to vapor, $\Delta \bar{H}_{\ell v}$, is nearly constant with respect to temperature near the boiling temperature and since $\bar{V}_v \gg \bar{V}_\ell$, then the integral of the Clausius–Clapeyron equation is a good approximation, namely

$$p_{o\ell v} = C_{\ell v} e^{-\dfrac{\Delta \bar{H}_{\ell v}}{RT}}, \tag{50}$$

where $C_{\ell v}$ is a constant of integration and $p_{o\ell v}$ is the vapor pressure of the pure solvent subject to its own vapor pressure. Since a tension is induced in the solvent in the solution to change its pressure from $p_{o\ell v}$ on the pure solvent to $p_\ell$ such that $(p_\ell - p_{o\ell v}) = -\Pi$, this changes the vapor from $p_{o\ell v}$ to

$$p_{\ell v} = p_{o\ell v} e^{\dfrac{\bar{V}_\ell}{RT}(p_\ell - p_{o\ell v})}, \tag{46}$$

according to Eq. (46) which was derived from the Poynting relation. Combining Eqs. (50) and (46), the vapor pressure of the solvent in a solution with an osmotic pressure $\Pi$ is then

$$p_{\ell v} = C_{\ell v} \, e^{-\left(\dfrac{\Delta \overline{H}_{\ell v} + \Pi \overline{V}_{\ell}}{RT}\right)} . \tag{51}$$

At the boiling temperature of the pure solvent, its vapor pressure

$$p_{o \ell v} = 1 \text{ atm} = C_{\ell v} \, e^{-\dfrac{\Delta \overline{H}_{\ell v}}{RT_b}} ; \tag{52}$$

and at the boiling temperature of the solution, its vapor pressure is

$$p_{\ell v} = 1 \text{ atm} = C_{\ell v} \, e^{-\left(\dfrac{\Delta \overline{H}_{\ell v} + \Pi \overline{V}_{\ell}}{RT_{\Pi b}}\right)} . \tag{53}$$

Combining these two Eqs. (52) and (53),

$$T_{\Pi b} - T_b = \frac{\Pi \overline{V}_{\ell}}{\Delta \overline{H}_{\ell v}} T_b \tag{54}$$

The osmotic pressure of an ideal solution at its boiling temperature is,

$$\Pi = \frac{RT_{\Pi b}}{\overline{V}_{\ell}} \frac{n_2}{n_1} . \tag{55}$$

Combining the last last two Eqs. (54) and (55), then

$$T_{\Pi b} = \frac{T_b}{1 - \dfrac{RT_b}{\Delta \overline{H}_{\ell v}} \dfrac{n_2}{n_1}} \tag{56}$$

For example, for an osmotic concentration of 1 mole of an ideal solute in 55.5009 moles of $H_2O$ (i.e. 1000 ml at $0°$ contains 55.5009 moles = $\dfrac{999.868 \text{ g}}{18.01534 \text{ g/mol}}$) for which

$$\Delta \overline{H}_{\ell v} = 4.0626 \cdot 10^4 \text{ J/mol}$$

and

$$T_b \equiv 273.1696 + 100 = 373.1696 \text{ K}$$

$$T_{\Pi_b} = 373.6838$$

and

$$\Delta T_{\Pi_b} = T_{\Pi_b} - T_b = 0.5142 \text{ K} .$$

The boiling point is raised by $0.5142°C$ by one mole of solute in 1000 ml of water; and the raising of the boiling temperature, like other colligative properties, is attributed to the enhanced tension in the water induced by the thermal pressure of the solute molecules exerted against the free surface of the solution.

## Experimental Confirmation of Relationship between Osmotic Pressure and Lowering of Vapor Pressure

All the colligative properties of any solution for all concentrations are derived from a common cause which is the greater tension in the solvent of the solution, $\Delta\tau_1$, than in pure solvent. An essential corollary of this assertion is that precise thermodynamic relationships can be derived relating each of the colligative properties to the other colligative properties under all conditions. If $n_2$ moles of solute molecules, ions, molecular complexes or particles are dissolved or dispersed throughout $n_1$ moles of free solvent molecules which do not further solvate with the solute, then the osmotic pressure of the solution will be

$$\Pi = \frac{RT}{\overline{V}_1} \frac{n_2}{n_1} \tag{57}$$

when the molecular weight and concentration of the solute are not large. For large solute molecules in high concentrations the pressure they exert against the free surface of the solution may exceed $\frac{RT}{\overline{V}_2} \frac{n_2}{n_1}$ by an additional matrix pressure $\Pi_m$. The effective osmotic pressure $\Pi_e$, and the enhanced tension induced in the solvent, $\Delta\tau_1$ of the solution are then

$$\Delta\tau_1 = \Pi_e = \Pi_T + \Pi_M = \frac{RT}{\overline{V}_1} \frac{n_2}{n_1} + \Delta\tau_M , \tag{58}$$

where $\Pi_T + \Pi_M$ are the thermal and mechanical components of $\Pi_e$ and where the magnitude of $\Delta\tau_M = \Pi_M$ . The magnitude of $\Pi_M$ and its dependence on molecular size and concentration are most reliably dealt with by obtaining empirical data.

Whatever the magnitude of $\Pi_e$, since $\Delta\tau_1$ always equals $\Pi_e$, there will be a precise thermodynamic relationship between $\Pi_e$ and the vapor pressure depression, $\Delta p_v$, the melting point lowering, $\Delta T_{\Pi_{e_m}}$, and the boiling point increase $\Delta T_{\Pi_{e_b}}$. The relationship between $\Pi_e$ and $\Delta p_v$ is derived from the Poynting relation

$$\int_o^{\Pi_e} \overline{V}_\ell(p)\, dp_\ell = \int_{p_{v_o}}^{p_v} \overline{V}_v dp_v \; . \tag{59}$$

To evaluate the liquid term of the Poynting relation for a compressible solution, the dependence of the molar volume of the solvent on pressure, $\overline{V}_\ell(p)$, must be known. The compressibility $\kappa$ is defined as

$$\kappa = - \frac{d\ell n \overline{V}_\ell}{dp_\ell} ,$$

so that $\overline{V}_\ell(p) = \overline{V}_\ell(o)\, e^{-\kappa p_\ell}$. The liquid term of the Poynting relation becomes

$$\overline{V}_\ell(o) \left|_a^{a-\Pi_e} e^{-\kappa p_\ell}\, dp_\ell \right. = \frac{\overline{V}_\ell(o)\, e^{-\kappa a}}{\kappa}(1 - e^{\kappa \Pi_e})$$

where $a$ is 1 atm.

Assuming the vapor of the solvent obeys the ideal gas law, then to an excellent approximation, the right term of the Poynting relation is $+RT\, \ell n\; p_v/p_{v_o}$ and the relationship between osmotic pressure and $\Delta p_v$ becomes

$$\ell n\, \frac{p_v}{p_{v_o}} = \frac{\overline{V}_\ell(o) e^{-\kappa a}}{\kappa RT}(1 - e^{\kappa \Pi_e}) \; . \tag{60}$$

If the solvent is nearly incompressible, i.e. $\kappa$ is very small, then this relationship reduces to

$$\Delta p_v/p_v = e^{-\frac{\Pi_e \overline{V}}{RT}} \; ;$$

and if matrix pressure, $\Pi_M$, exerted by the solute molecules is negligible, then

$$\Delta p_v/p_v = e^{-\frac{n_2}{n_1}}.$$

For an incompressible solution with negligible matrix pressure, the melting point depression may be derived from Eq. (48) as

$$\Delta T_{\Pi_m} = T_{\Pi_m} - T_m = -\frac{\Pi \bar{V}_1 T_m}{\Delta \bar{H}_{s\ell}}$$

and the boiling point elevation may be derived from Eq. (54) as

$$\Delta T_{\Pi_b} = T_{\Pi_b} - T_b = \frac{\Pi \bar{V}_1 T_b}{\Delta \bar{H}_{\ell v}}.$$

Experimental evidence for the precise relationship between osmotic pressure and vapor pressure lowering is the remarkable data by Berkeley et al. (1909, 11). They measured directly both the vapor pressure ratio, $p_v/p_{vo}$, and the osmotic pressure, $\Pi_e$, for several concentrations of calcium ferrocyanide solutions ranging from 31 to 50 g of anhydrous salt in 100 g of water at $0°C$. They also measured the compressibility $\kappa$, of these solutions at each concentration. Their experimental data are presented in columns 1 to 4 in Table 1. They enable us to calculate the ratio of the vapor pressure of the solution to the vapor pressure of pure water by the thermodynamic Eq. (60) and the results are given in column 5. The experimental values of $p_v/p_{vo}$ exceed pres-theoretical values by less than 0.4% over a wide range of osmotic pressures measured from 40 to 130 atm. At these very high concentrations, the measured osmotic pressures greatly exceeds pressures induced by Brownian motion alone. We have also computed the thermal component of the osmotic pressure, $\Pi_T$, by Eq. (57) which is given in column 9. The matrix component, $\Pi_M = \Pi_e - \Pi_T$, is shown in column 10. Notwithstanding the fact that the measured osmotic pressure of a concentrated solution may greatly exceed the thermal osmotic pressure, the actual osmotic pressure of the solution and the lowering of its vapor pressure must have a common cause. The common cause is the increased tension in the solvent of the solution induced by thermal and mechanical pressures exerted by the solute on the free surface of the solution.

Table 1

| 1 | 2 | 3 | 4 | 5 | 6 | 7 | 8 | 9 | 10 |
|---|---|---|---|---|---|---|---|---|---|
| $gCa_2Fe(CN)_6$ per 100 g water [a] | $\Pi_e$ atm | $p_v/p_{v_o}$ (measured) | $\kappa$ $10^5$ $atm^{-1}$ | $p_v/p_{v_o}$ (calculated) [b] | $\dfrac{meas.\,p_v/p_{v_o}}{calc.\,p_v/p_{v_o}}\cdot 100$ | $n_2$ mol [c] | $n_1$ mol | $\dfrac{RT}{V_1}\dfrac{n_2}{n_1}$ atm | $\Pi_M$ atm |
| 49.966 | 131.00 | 0.903261 | 2.270 | 0.900082 | 100.35 | 0.17105 | 3.66939 | 58.10 | 72.6 |
| 47.219 | 112.84 | 0.916028 | 2.368 | 0.913326 | 100.30 | 0.16165 | 3.77283 | 48.51 | 64.3 |
| 42.889 | 87.09 | 0.934431 | 2.510 | 0.932436 | 100.21 | 0.14682 | 3.93588 | 42.03 | 45.1 |
| 39.503 | 70.84 | 0.946002 | 2.615 | 0.944695 | 100.14 | 0.13523 | 4.06339 | 41.48 | 29.4 |
| 31.388 | 41.22 | 0.967904 | 2.955 | 0.967448 | 100.05 | 0.10745 | 4.36897 | 30.65 | 10.6 |

[a] anhydrous salt; Mol Wt = 292.115 g/mol

[b] calculated by Eq. (60)

[c] calculated on the basis that each mole of anhydrous salt hydrates with 11 mol water in aqueous solution. By drying an aqueous solution of $Ca_2Fe(CN)_6$, Berkeley et al., determined the water of crystallization to be 16.59 percent by weight or 11 mol water per mol anhydrous salt (11)

In a Solution

The chemical potential of the solvent in a solution may be altered by changing the temperature and external pressure applied
to the solution and by changing the mole fraction of the solute
in the solution. For a homogeneous solution the change in chemical potential of the solvent is expressed as

$$d\mu_1 = -\bar{S}_1 dT + \bar{V}_1 dp + \frac{\partial \mu_1}{\partial x_2} dx_2 \qquad (5)$$

a thermodynamic statement which was derived as Eq. (5) in Chapter II. If the temperature and the externally applied pressure
are held constant while changing the mole fraction from $x_2 = 0$
for pure solvent to $x_2 = \frac{n_2}{n_1 + n_1}$ for the solution, then the total change in the chemical potential of the solvent for this
homogeneous solution becomes

$$\Delta\mu_1 = \int d\mu_1 = \int_{x_2 = 0}^{x_2 = \frac{n_2}{n_1 + n_2}} \frac{\partial \mu_1}{\partial x_2} dx_2 \ .$$

This thermodynamic statement provides no clue concerning the
alteration in the solvent which alters its chemical potential.
For many purposes an explanation is not required. On the other
hand, to account for the change in chemical potential of a solution formed at constant temperature and external pressure,
we need only recognize that the tension in the solvent has been
enhanced by

$$\Delta\tau_1 = \frac{RT}{\bar{V}_1} \frac{n_2}{n_1} + \Delta\tau_M \ .$$

This applies to an ideal solution when the mole fraction of the
solute changes from 0 to $\frac{n_2}{n_1 + n_2}$ in the solution. This enhanced tension, induced by the thermal pressure and the mechanical pressure
of the solute molecules against the unrestrained surface of the

solution, lowers the chemical potential of the solvent in the solution with respect to pure solvent.

The third term on the right in Eq. (5) states that the chemical potential of the solvent changes, at constant temperature and external pressure, by an amount which equals the rate of change of the chemical potential of the solvent with respect to the change in the mole fraction times the change in the mole fraction while forming the solution. Since this statement has no inherent mechanistic meaning, it can readily be interpreted to mean that the change, in chemical potential of the solvent is attributable to the negative hydrostatic pressure that was internally induced in the solvent, that is

$$\Delta\mu_1 = \int_0^{\frac{n_2}{n_1 + n_2}} \frac{\partial\mu_1}{\partial x_2}dx_2 = -\bar{V}_1\Delta\tau_1 = -RT\frac{n_2}{n_1} - \bar{V}_1\Delta\tau_M \, ,$$

at constant temperature and external pressure.

The enhanced tension in the solvent also lowered its vapor pressure in accordance with Eq. (59). Thus, the change in chemical potential in the solvent may be related to the lowering of the vapor pressure of the solvent in the solution, that is

$$\Delta\mu_1 = RT \ln p_V/p_{V_O} \, .$$

Similarly, the lowering of the chemical potential of the solvent may be related to the osmotic pressure and other colligative properties of the solution. However, it seems obvious to us that all the alterations in the solution, including the lowering of its chemical potential, are directly attributable to the enhanced tension in its solvent.

## In a Matrix

A submerged wettable matrix exerts no thermal pressure on the fluid which it contains. Nevertheless, when the boundaries of the fluid and the matrix coincide, the matrix can exert a force against the boundaries of a fluid. A substance possessing these properties may be constituted of long chain fibrous molecules, such as cellulose, which are wetted by water. The cell wall of a plant cell is a matrix constituted of molecules of cellulose and hemicellulose which are cross-linked with lignin and pectin. The cell wall matrix can exert many atmospheres of pressure against the boundary of the water and induce an opposing tension in the water which permeates the matrix. When the liquid extends beyond the boundary of the matrix, the matrix exerts no pressure on the boundary of the liquid and induces no tension in the liquid. As the liquid evaporates and recedes to a boundary less than the boundary of the matrix, the wettable fibers of the matrix adhere to the liquid, exerting a force on the liquid and inducing an opposing tension in the liquid. The tension in the liquid lowers the vapor pressure and may be said to lower the chemical potential of the liquid.

A change in the molar Gibbs free energy of the liquid, as its tension increases is

$$\Delta\mu_1 \equiv \Delta\bar{G}_1 = -\bar{S}_1\Delta T - \bar{V}_1\Delta\tau_1 \ .$$

If the temperature is held constant, then the change in the chemical potential becomes

$$\Delta\mu_1 = -\bar{V}_1\Delta\tau_1$$

where $\Delta\tau_1$ is the negative hydrostatic pressure induced in the liquid by the mechanical force exerted by the matrix as the liquid evaporates. The negative hydrostatic pressure (hydrostatic tension) in the liquid lowers its vapor pressure in accordance with a thermodynamic relationship, the Poynting relation, that is

$$-\bar{V}_1\Delta\tau_1 = RT\ln p_v/p_{v_o}$$

where $p_{v_o}$ is the vapor pressure of the liquid without the matrix pressure and $p_v$ is the vapor pressure when the matrix induced a tension $\Delta\tau_1$. Thus the tension in the liquid lowers its vapor pressure by

$$\Delta p_v = p_v - p_{v_o} = (e^{-\frac{\bar{V}_1\Delta\tau_1}{RT}} - 1)\,p_{v_o} \ .$$

We may, if we choose, also relate the change in chemical potential to the lowering of the vapor pressure; thus,

$$\Delta\mu_1 = RT\ln p_v/p_{v_o} \ .$$

Of course, the induced tension in the liquid lowers its chemical potential and its vapor pressure; so these effects must be relatable. However, we should not suppose that lowering the vapor pressure "causes" the lowered chemical potential.

The matrix of fibers will be able to induce an opposing tension in the liquid which permeates it only if the spacing between the fibers is very small. For the enhanced tension in the liquid to attain a value $\Delta\tau$, the spacings between all fibers must be equal to or less than $d$ where $d$ is given by

$$\Delta\tau = \frac{4\sigma}{d}$$

and where $\sigma$ is the surface tension of the liquid. For spacings greater than $d$, the liquid-vapor interface will recede inward from the boundary of the matrix when the enhanced tension equals $\Delta\tau$ as the liquid evaporates.

If an edge of a dry vertical matrix is immersed in a pool of liquid which wets the fibers of the matrix, the liquid will rise in the matrix against gravity, drawn upward by capillarity, i.e. the adhesive forces between the fibers of the matrix and the liquid. The matrix is said to imbibe the liquid and the process of liquid entering the matrix is known as imbibition. Again, the distance the liquid will rise in the matrix will depend upon the spacing between the fibers, as in a capillary tube. The tension at the upper surface of the liquid will be $\rho g h$, where $\rho$ is the density of the liquid and $h$ is the height to which the liquid rises in the matrix. The tension in the liquid at height $h$ will lower the vapor pressure of the liquid at $h$ such that

$$\ln\left(1 + \frac{\Delta p_V}{p_{V_O}}\right) = \frac{\overline{V}_1 \rho g h}{RT} .$$

Indeed, the Second Law of Thermodynamics requires that the vapor pressure in the liquid in the matrix at any elevation $z$ above the level of the pool of liquid be under an enhanced tension $\Delta\tau(z) = \rho g z$. Only then can the vapor pressure in the liquid in the matrix at $z$ exactly match the Boltzmann distribution of vapor above the pool of liquid at equilibrium in a convection free enclosure.

## Chemical Potential of Solvent in Real Solutions

A dilute solution may be constituted of a solute and solvent for which the solute does not dissociate in, nor solvate with, the solvent and for which the partial molar volumes of solute and solvent equal to their respective molar volumes. The chemical potential and the vapor pressure of the solvent in this solution is precisely stated by the formal logic of thermodynamics. Since the vapor pressure of this solution may differ from the vapor pressure of most real solutions of equal concentration, the elegant logic of thermodynamics may be retained by one of two procedures. 1. We may recognize that the chemical potential of the solvent in a real solution is lower than in pure solvent because an enhanced tension has been induced in the solvent, attributable to thermal and mechanical pressures by the solute. This enhanced tension is, as derived in Eq. (58)

$$\Delta\tau_1 = \frac{RT}{\overline{V}_1} \frac{n_2}{n_1} + \Delta\tau_M$$

where we understand that $n_2$ is the number of moles of solute (subunits, if it dissociates or complex units, if it solvates), $n_1$ is the number of moles of free solvent, $\overline{V}_1$ is the molar volume of pure solvent and $\Delta\tau_M$ is the enhanced solvent tension attributable to the mechanical or matrix pressure of the solute molecules, $\Pi_M$. By this procedure,

$$\Delta\mu_1 = -\bar{V}_1\Delta\tau_1$$

and

$$\Delta\mu_1 = RT\ln p_{v_1}/p_{v_0} ,$$

for constant temperature and external pressure. 2. The classical procedure has been to define a standard solution as one which is described by Raoult's Law,

$$p_{v_1}/p_{v_0} = \frac{n_1}{n_1 + n_2} = x_1 .$$

If the ratio of the vapor pressure of the solvent in a real solution to the vapor pressure of its pure solvent deviates from Raoult's Law and is not equal to the mole fraction of solvent in the solution, then a new term, fugacity, is defined for the solution such that the ratio of the fugacity of the solvent in a real solution to the fugacity of its pure solvent does obey Raoult's Law, i.e.

$$f_1/f_0 \equiv x_1 .$$

In the limit

$$\lim_{x_1 \to 1} f_1/f_0 = p_{v_1}/p_{v_0} .$$

By definition, the change in chemical potential of the solvent for the standard solution is

$$\Delta\mu_1 = RT\ln p_{v_1}/p_{v_0} = RT\ln x_1 ;$$

and by definition, the change in chemical potential of the solvent in a real solution is given by

$$\Delta\mu_1 = RT\ln x_1$$

where $x_1 \equiv f_1/f_0$. For the real and standard solutions, the change in chemical potential of the solvent has no mechanistic meaning since it means only

$$\Delta\mu_1 = \int_0^{x_2} \frac{\partial\mu_1}{\partial x_2}dx_2 ,$$

at constant temperature and constant external pressure. For most users of this procedure there is an irresistable inclination to presume a cause and effect relationship between $p_{v_1}/p_{v_0}$ and $\Delta\mu_1$, or between $f_1/f_0$ and $\Delta\mu_1$.

Our final conclusion is that both the thermodynamic and kinetic treatment of the colligative properties of a solution are well served by recognizing that the solute enhances the tension in the solvent relative to the tension in pure solvent. Thus, there is no need to define fugacities or any other coefficients to deal with the colligative properties of real solutions.

# Glossary of Terms

| | |
|---|---|
| $A$ | cross sectional area, cm$^2$ or m$^2$ |
| $a$ | atmospheric pressure; 1 atm = $1.01325 \cdot 10^6$ dyn/cm$^2$ |
| $B$ | cross sectional area, cm$^2$ or m$^2$ |
| bar | pressure unit = $10^6$ dyn/cm$^2$; 1 atm = 1.01325 bar |

$C$      concentration: may be defined in three ways.

1. The *average molar concentration of the solute* is the number of moles per liter of solution, i.e. $\bar{C}_2 = \frac{n_2}{V_S}$ . Similarly, the *average molar concentration of the solvent* may be defined $\bar{C}_1 = \frac{n_1}{V_S}$; where $V_S$ is the volume of the solution in liters.

2. The *average tensile concentration of the solute* is the number of moles of solute per liter of solvent, i.e. $\bar{C}_2' = \frac{n_2}{N_1 v_1} = \frac{n_2}{n_1 V_1}$. Similarly, the *average tensile concentration of the solvent* may be defined, $\bar{C}_1' = \frac{n_1}{N_1 v_1} = \frac{n_1}{n_1 V_1} = \frac{1}{V_1}$, where $\bar{V}_1$ is the partial molar volume of the solvent in liters.

3. The *average molal concentration of the solute* is the number of moles of solute per kilogram of solvent, i.e. $\bar{C}_2'' = \frac{n_2}{N_1 m_1} = \frac{n_2}{n_1 M_1}$. Similarly the *average molal concentration of the solvent* may be defined $\bar{C}_1'' = \frac{n_1}{N_1 m_1} = \frac{1}{N_A m_1} = \frac{1}{M_1}$, where $M_1$ is the molecular weight of the solvent in kilograms. Only if the density of the solvent is 1 g/cm$^3$ will the osmotic and molal concentrations be equal. Only the molal concentration is independent of temperature so that only at $3.98^\circ$C are the osmotic and molal concentrations of a water solution exactly equal.

$c_0(o)$      molar concentration of solvent molecules at the surface of pure solvent

$c_1(z)$        molar concentration of solvent molecules in solution at height $z$

$c_2(z)$        molar concentration of solute molecules in solution at height $z$

$c_1'(z)$        tensile concentration of solvent molecules in solution at height $z$

$c_2'(z)$        tensile concentration of solute molecules in solution at height $z$

$E$        energy is an extensive function of the state of a system

Extensive Function        functions of the state of the system that depend upon the number of moles and species of molecules, $n_i$; e.g. $V$, $G$ and $S$

$G$        Gibbs free energy of a system: $G = H - TS = E + pV - TS$.

$G(T,p,n_1,n_2,...)$ the Gibbs function is an extensive function of the state of a system and depends upon the number of moles and species of molecules, $n_1$, as well as upon temperature and pressure

$\overline{G}_i$        partial molar Gibbs free energy of species $i$

$g$        acceleration due to gravity; 1 g $= 980.7 \pm 0.9$ cm/s$^2$

$\Delta \overline{H}_{\ell v}$        partial molar enthalpy for transition from liquid to vapor (heat of vaporization) $= 4.0626 \cdot 10^4$ J/mol for water to steam

$\Delta \overline{H}_{s \ell}$        partial molar enthalpy for transition from solid to liquid (heat of fusion) $= 5.983 \cdot 10^3$ J/mol for ice to water

$\Delta \overline{H}_{sv}$        partial molar enthalpy for transition from ice to vapor (heat of sublimation) $= 4.6609 \cdot 10^4$ J/mol for ice to steam

Ideal solution        $V_S = V_1 + V_2$, the volume of the solution equals the separate volumes of solvent and solute

| | |
|---|---|
| Intensive function | functions of the state of the system that are independent of the number of moles and species of molecules, $n_i$; e.g. $T$, $p$, $\bar{V}_i$, $\bar{G}_i$, $\mu_i$, and $\bar{S}_i$ |
| $k$ | Boltzmann constant = $1.38042 \cdot 10^{-23}$ J K$^{-1}$ |
| $M_i$ | molecular weight of molecular species $i$ |
| $M_w$ | molecular weight of water = 18.01534 g/mol |
| $m_i$ | mass of molecule of species $i$ |
| $N_A$ | Avogadro's number = $6.023 \cdot 10^{23}$ molecules per gram molecular weight of any species |
| $N_1$ | number of molecules of solvent in a solution |
| $N_2$ | number of molecules of solute in a solution |
| $n_1$ | number of moles of solvent in a solution |
| $n_2$ | number of moles of solute in a solution |
| $dn_1(z)$ | number of moles of solvent in a solution between a vertical height $z$ and $z + dz$ |
| $dn_2(z)$ | number of moles of solute in a solution between a vertical height $z$ and $z + dz$ |
| $n_i$ | number of moles of species $i$ in a solution |
| $p$ | pressure is an intensive function of the state of a system |
| $p_i(x_i)$ | thermal pressure exerted by molecular species $i$ when its mole fraction in the solution is $x_i$ |
| $p_\ell$ | external pressure applied to a liquid. Subscripts v, $\ell$, s, designate vapor, liquid and solid. |
| $p_{o\ell}$ | initial pressure applied to liquid |
| $p_{o\ell v}$ | vapor pressure of a liquid subject to its own vapor pressure |
| $p_{ov}$ | initial vapor pressure of liquid |
| $p_{ovm}$ | vapor pressure of pure liquid at its melting temperature and subject to its own vapor pressure |
| $p_{osv}$ | vapor pressure of pure solid subject to its own vapor pressure |

| | |
|---|---|
| $p_V$ | vapor pressure of liquid |
| $p_{V0}$ | vapor pressure of a pure liquid subject to its own vapor pressure |
| $p_0$ | thermal pressure exerted by the Brownian motion of the solvent molecules at the unrestrained surface of pure solvent |
| $p_1(h)$ | thermal pressure exerted by the Brownian motion of solvent molecules in a solution at the unrestrained surface at height $h$ |
| $p_2(h)$ | thermal pressure exerted by the solute molecules at the surface at height $h$ |
| $R$ | universal gas constant $= kN_A = 8.3143$ J mol$^{-1}$ K$^{-1}$ |
| $S$ | entropy is an extensive function of the state of a system; $dS = \dfrac{\delta Q}{T}$ |
| $T$ | temperature is an intensive function of the state of a system, $K$ |
| $T_b$ | boiling temperature of pure solvent at 1 atm |
| $T_m$ | melting temperature of pure solvent at 1 atm |
| $T_{IIm}$ | triple point of water: the temperature that ice and water have equal vapor pressures at the vapor pressure of water; $= 0.0098°C = 273.1696$ K |
| $V$ | volume is an extensive function of the state of the system |
| $V_S$ | volume of a solution in liters |
| $V_1$ | volume of solvent to be added to solution in liters |
| $V_2$ | volume of solute to be added to solution in liters |
| $\overline{V}_i$ | partial molar volume of molecular species $i$ |
| $\overline{V}_\ell$ | molar volume of liquid in liters/mole |
| $\overline{V}_s$ | molar volume of solid in liters/mole |
| $\overline{V}_v$ | molar volume of vapor in liters/mole |
| $\overline{V}_1$ | partial molar volume of solvent in liters/mole |

$\overline{V}_2$        partial molar volume of solute in liters/mole

$v_i$        volume of molecule of species $i$

$x_i$        mole fraction of species $i$ in solution;

$$x_i = \frac{n_i}{\sum\limits_j n_i}$$

$0°C$        zero degree Celsius, the temperature at which ice and water have equal vapor pressures at 1 atmosphere pressure = 273.1598 K

$\mu_i$        chemical potential of the $i$th component of the system equal to $\overline{G}_i$ by definition

$\Pi$        osmotic pressure of a solution, the pressure applied to an enclosed solution sufficient to prevent entry of solvent through the semipermeable membrane. $-\Pi$ is the negative pressure applied to the pure solvent sufficient to prevent its entry through the membrane into the solution

$\rho_i$        density of species $i$

$\tau_1$        tension in solvent in solution opposing the thermal pressure of solute and solvent molecules exerted at the unrestrained surface of the solution

$\tau_0$        tension in pure solvent opposing the thermal pressure of solvent molecules exerted at unrestrained surface of the pure solvent

# References

1. Apfel, R.E.: The tensile strength of liquids. Sci. Am., Dec., 58-71 (1972).
2. Arrhenius, S.: Recherches sur la conductibilité galvanique des électrolytes, I and II. Bih. K. Vet. Akad. Handl. $\underline{8}$ (13 and 14) (1884).
3. Arrhenius, S.: Über die Dissociation der in Wasser gelösten Stoffe, Z. Phys. Chem. $\underline{1}$, 631-648 (1887).
4. Arrhenius, S.: Einfache Ableitung der Beziehung zwischen osmotischem Druck und Erniedrigung der Dampfspannung. Z. Phys. Chem. $\underline{1}$, 115-119 (1889).
5. Asimov, I.: Asimov's Biographical Encyclopedia of Science and Technology. Garden City, N.Y.: Doubleday and Co. (1972).
6. Askenasy, E.: Über das Saftsteigen. Verh. Naturhist. Med. Ver. Heidelberg, N.F. $\underline{5}$, 325-345 (1895).
7. Askenasy, E.: Beiträge zur Erklärung des Saftsteigens. Verh. Naturhist. Med. Ver. Heidelberg $\underline{5}$, 429-448 (1896).
8. Bailey, I.W.: The structure of the bordered pits of conifers and its bearing upon the tension hypothesis of the ascent of sap in plants. Bot. Gaz. $\underline{62}$, 133-142 (1916).
9. Bennett, J.P., Anderssen, F.G., Milad, Y.: Methods of obtaining tracheal sap from woody plants. New Phytologist $\underline{26}$, 316-323 (1927).
10. Berkeley, Earl of, Hartley, E.G.: Dynamic osmotic pressures. Proc. Roy. Soc. Ser. A. $\underline{82}$, 271-275 (1909).
11. Berkeley, Earl of, Hartley, E.G., Burton, C.V.: On the osmotic pressures of aqueous solutions of calcium ferrocyanide. Part I. Concentrated solutions. Phil. Trans. A $\underline{209}$, 177-203 (1909).
12. Berthelot, P.M.: Sur quelques phénomènes de dilatation forcée des liquides. Ann. Chim. Phys. $\underline{3}$, 30, 232-242 (1850).
13. Boyer, J.S.: Matric potentials of leaves. Plant Physiol. $\underline{42}$, 213-217 (1967).
14. Briggs, L.: Limiting negative pressure of water. J. Appl. Physics. $\underline{21}$, 721-722 (1950).
15. Callendar, H.L.: On vapour-pressure and osmotic pressure of strong solutions. Proc. Roy. Soc. A $\underline{80}$, 466-500 (1908).
16. Dainty, J.: Water relations of plant cells. Adv. Bot. Res. $\underline{1}$, 279-326 (1963).
17. DeVries, H.: Ueber die Anziehung zwischen Gelösten Stoffen und Wasser in verdünnten Lösungen. Verslagen en Mededeelingen der Koninklijke Akademie van Wetenschappen, Afdeeling Natuurkunde, 2de Reeks, Deel XIX:314 (1883).
18. DeVries, H.: Sur la force osmotique des solutions diluées. Compt. Rend. 97, 1083 (1883).
19. DeVries, H.: Eine Methode zur Analyse der Turgorkraft. Pringsheims Jb. Wiss. Bot. $\underline{14}$, 427 (1884).
20. DeVries, H.: Osmotische Versuche mit lebenden Membranen. Phys. Chem. $\underline{2}$, 415 (1888).

21. Dixon, H.H.: Transpiration and the Ascent of Sap in Plants, 1-216. London, Macmillan and Co. (1914).
22. Dixon, H.H., Joly, J.: On the ascent of sap. Ann. Bot. $\underline{8}$, 468-470 (1894).
23. Duclaux, J.: Théorie de gas, J. Chim. Phys. $\underline{65}$, 435-438 (1965).
24. Duhem, P.: De l'influence de la pesanteur sur les dissolutions. J. Phys. Theorique et appliquée, 7, 391-419 (1888).
25. Dutrochet, H.: Mémoires des végétaux et des animaux. I. De L'endosmose. Bruxelles: Meline Cans et Co. (1837).
26. Edwards, I.E.S.: The Pyramids of Egypt. Baltimore, Maryland: Penguin Books (1964).
27. Findlay, A.: Osmotic Pressure. Mon. Inorg. Phys. Chem. London: Longmans, Green and Co. (1913).
28. Garby, L.: Studies of transfer of matter across membranes with special reference to the isolated human amniotic membrane and the exchange of amniotic fluid. Acta. Physiol. Scand. $\underline{40}$ (Suppl. 137), 89 (1957).
29. Glasstone, S., Lewis, D.: Elements of physical chemistry. Princeton, N.J.: D. Van Nostrand Co. (1960).
30. Guldberg, C.M.: Sur la loi des points de congélation de solutions salines. Compt. Rend. $\underline{70}$, 1349-1352 (1870).
31. Gouy, M., Chaperon, G.: Sur l'équilibre osmotique. Ann. Chim. et Phys. $\underline{6}$ (13), 120-132 (1888).
32. Hales, S.: Vegetable Staticks. Sci. Book Guild. London: Beaverbrook (1727).
33. Hamburger, H.J.: Proc. K. Akad. Wetensch. Amsterdam (1883). Ref. Findlay.
34. Hammel, H.T.: Freezing of xylem sap without cavitation. Plant. Physiol. $\underline{42}$, 55-66 (1967).
35a. Hammel, H.T.: Letters to the editors: On the ascent of sap. Science $\underline{179}$, 1248-1249 (1973).
35b. Hammel, H.T.: Colligative properties of a solution. Enhanced tension in solvent explains alterations of solution. Science (in press).
36. Hammel, H.T., Scholander, P.F.: Thermal motion and forced migration of colloidal particles generate hydrostatic pressure in solvent. Proc. Nat. Acad. Sci., USA $\underline{70}$, 124-128 (1973).
37. Hargens, A.: Thesis (1972). (See Scholander, P.F., 1967. In Letters. Science $\underline{158}$, 1212.
38. Hargens, A., Scholander, P.F.: Stretch mounting of osmotic membrane. Micro. Vasc. Res. $\underline{4}$, 417-419 (1969).
39. Heezen, B.C., Ewing, M.: Turbidity currents on submarine slumps and the 1929 Grand Banks earthquake. Am. J. Science $\underline{250}$, 849-873 (1952).
40. Herzfeld, K.F.: Thermodynamische und kinetische Betrachtungen über die Zustandekommen der Dampfdruckerniedrigung von Lösungen. Phys. Z. $\underline{38}$, 58-64 (1937).
41. Hulett, G.A.: Beziehung zwischen negativem Druck und osmotischem Druck. Z. Phys. Chem. $\underline{42}$, 353-368 (1902).
42. Kedem, O., Katchalsky, A.: Thermodynamic analysis of the permeability of biological membranes to non-electrolytes. Biochim. et Biophys. Acta. $\underline{27}$, 229 (1958).
43. Levitt, J., Storvick, T.S.: Letters. Science $\underline{179}$, 1250 (1973).
44. Lewis, G.N., Randall, M.: Thermodynamics. Revised by Pitzer, K.S. and Brewer, L., Ed. II. New York: McGraw-Hill (1961).

45. Lloyd, D.J., Moran, T.: Pressure and the water relations of proteins through isoelectric gelatin gels. Proc. Roy. Soc. 7(A), 382-395 (1934).
46. Loeb, L.B., Adams, A.S.: The Development of Physical Thought. New York: Wiley and Sons, Inc. (1933).
47. MacDougal, D.T.: Reversible variations in volume, pressure, and movements of sap in trees. Carnegie Inst. Wash. Publ. 365 (1925).
48. Mauro, A.: Nature of solvent transfer in osmosis. Science 126, 252 (1957).
49. Mauro, A.: Some properties of ionic and nonionic semipermeable membranes. Circulation 21, 845-854 (1960).
50. Mauro, A.: Osmotic flow in a rigid porous membrane. Science 149, 867-869 (1965).
51. Mees, G.C., Weatherley, P.L.: The mechanism of water absorption by roots. Proc. Roy. Soc. B147, 367-381 (1957).
52. Meschia, G., Setnikar, I.: Experimental study of osmosis through a collodion membrane. J. Gen. Physiol. 42, 429-444 (1958).
53. Meyer, L.: Über das Wesen des osmotischen Druckes. Z. Phys. Chem. 5, 23-27 (1890).
54. Mordy, W.A.: Computations of the growth by condensation of a population of cloud droplets. Tellus II:1-19 (1959).
55. Mysels, K.: Introduction to Colloid Chemistry. New York: Interscience Publishers, Inc. (1959).
56. Noyes, A.: Die genaue Beziehung zwischen osmotischem Druck und Dampfdruck, Z. Phys. Chem. 35, 707-721 (1900).
57. Oertli, J.J.: The stability of water under tension in the xylem. Z. Pflanzenphysiol. 65, 195-209 (1970).
58. Ogston, A.G.: On water binding. Fed. Proc. 25, 986-989 (1966).
59. Pappenheimer, J.R.: Passage of molecules through capillary walls. Physiol. Rev. 33, 389-423 (1953).
60. Perez, M., Scholander, P.F.: Molecular buoyancy and osmotic equilibrium. Proc. Nat. Acad. Sci. USA 69, 301-302 (1972).
61. Perrin, J.: Brownian movement and molecular reality. Ann. Chim. Phys. Ser. 8 (1909).(Transl. Soddy, F. in Taylor and Francis, London: 1910)
62. Pfeffer, W.: Osmotische Untersuchungen. 1-236. Leipzig: W. Engelman (1921).
63. Plesset, M.S.: Letters. Am. Scientist 61, 142 (1973).
64. Plumb, R.C., Bridgman, W.B.: Ascent of sap in trees. Science 176, 1129-1131 (1972).
65. Poynting, J.H.: Change of state: Solid-Liquid. Phil. Mag. 5, 32-48 (1881).
66. Poynting, J.H.: Osmotic Pressure, Phil. Mag. 42, 289-299 (1896).
67. Raoult, F.M.: Sur la tension de vapeur et sur le point de congélation de solutions salines. Compt. Rend. 87, 167-171 (1878).
68. Raoult, F.M.: Loi de congélation des solutions benzeniques des substances neutres. Compt. Rend. 95, 187 (1882).
69. Raoult, F.M.: Loi général de congélation des dissolvants. Compt. Rend. 95, 1030-1033 (1882).
70. Raoult, F.M.: Loi de congélation des solutions aqueuses des materières organiques. Ann. Chim. Phys. 28, 133-144 (1883).

71. Raoult, F.M.: Loi général des tensions de vapeur des dis-
    solvants. Compt. Rend. 104, 1430 (1887).
72. Ray, P.M.: On the theory of osmotic water movement.
    Plant Physiology 35, 783 (1960).
73. Renner, O.: Experimentelle Beiträge zur Kenntnis der Wasser-
    bewegung. Flora (Ger.) 103, 171-247 (1911).
74. Renner, O.: Versuch zur Mechanik der Wasserversorgung.
    I. Der Druck in den Leitungsbahnen von Freilandpflanzen.
    Ber. deut. bot. Ges. 30, 576-580 (1912).
75. Renner, O.: Zum Nachweis negativer Drucke im Gefässwasser
    bewurzelter Holzgewächse. Flora (Ger.), 118-119, 402-408
    (1925).
76. Riesenfeld, E.H.: Svante Arrhenius, (1859-1927). In: Grosse
    Männer, Bd. 11. Leipzig: Akad. Verl. Ges. (1931).
76b. Roedder, E.: Metastable superheated ice in liquid-water
    inclusions under high negative pressure. Science 155,
    1413-1417 (1967).
77. Scholander, P.F.: The role of solvent pressure in osmotic
    systems. Proc. Nat. Acad. Sci. 55, 1407-1414 (1966).
78. Scholander, P.F.: Osmotic mechanism and negative pressure.
    Science 156, 67-69 (1967).
79. Scholander, P.F.: How mangroves desalinate seawater.
    Physiol. Plantarum 21, 251-261 (1968).
80. Scholander, P.F.: Imbibition and Osmosis in Plants. Topics
    in the Study of Life, The Bio Source Book. New York:
    Harper and Row, Inc. (1971).
81. Scholander, P.F.: State of water in osmotic processes.
    Microvasc. res. 3, 215-232 (1971).
82. Scholander, P.F.: Tensile water. Am. Scientist 60, 585-590
    (1972).
82b. Scholander, P.F. Water states and water gates in osmotic
    processes, and the inoperative concept of molfraction of
    water, J. Exp. Zool. 194, 241-248 (1975).
83. Scholander, P.F., Bradstreet, E.D., Hammel, H.T., Hemming-
    sen, E.A.: Sap concentrations in halophytes and some other
    plants. Plant Physiol. 41, 529-532 (1966).
84. Scholander, P.F., Hammel, H.T., Bradstreet, E.D., Hemming-
    sen, E.A.: Sap pressure in vascular plants. Science 148,
    339-346 (1965).
85. Scholander, P.F., Hammel, H.T., Hemmingsen, E.A., Bradstreet,
    E.D.: Hydrostatic pressure and osmotic potential in leaves
    of mangroves and some other plants. Proc. Nat. Acad. Sci.
    USA 52, 119-125 (1964).
86. Scholander, P.F., Hammel, H.T., Hemmingsen, E.A., Garey, W.:
    Salt balance in mangroves. Plant Physiol. 37, 722-729 (1962).
87. Scholander, P.F., Hemmingsen, E.A., Garey, W.: Cohesive
    lift of sap in the rattan vine. Science 134, 1835-1838
    (1961).
88. Scholander, P.F., Love, W.E., Kanwisher, J.W.: The rise of
    sap in tall grapevines. Plant Physiology 30, 93-104 (1955).
89. Scholander, P.F., Perez, M.: Experiments on Osmosis with
    Magnetic Fluid. Proc. Nat. Acad. Sci. USA 68, 1093-1094
    (1971).
90. Scholander, P.F., Perez, M.: Effect of gravity on osmotic
    equilibria. Proc. Nat. Acad. Sci. USA 68, 1569-1571 (1971).
91. Sen-Gupta, J.: Die osmotischen Verhältnisse bei einigen
    Pflanzen in Bengal (Indien). Ber. deut. bot. Ges. 53,
    783-795 (1935).

130

92. Shepard, F.P.: Submarine Geology. 2nd ed. New York: Harper and Row (1963).

93. Slatyer, R.O.: Plant-water Relationsships. London: Academic Press (1967).

94. Staverman, A.J.: The theory of measurement of osmotic pressure. Recl. Trav. Chim. Pays-Bas. Belg. 70, 344-352 (1951).

95. Strasburger, E.: Über den Bau und die Verrichtungen der Leitungsbahnen in den Pflanzen. Hist. Beitr. Eduard Strasburger 3, 1-1000 (1891)

96. Strasburger, E.: Über das Saftsteigen, Hist. Beitr. Eduard Strasburger 5, 1-94 (1893).

97. Sullivan, E.C.: Biographical Memoirs. Nat. Acad. Sci., George Augustus Hulett 34, 83-105 (1960).

98. Thomson, W.: On the equilibrium of vapour at a curved surface of liquid. Phil. Mag.(A) 42, 448-452 (1871).

99. Traube, M.: Arch. Anat., Physiol., Wiss. Med. (1867), quoted Findlay 1913.

100. Tyree, M.T., Dainty, J., Benis, M.: The water relations of hemlock *(Tsuga canadensis)*. I. Some equilibrium water relations as measured by the pressure-bomb technique. Can. J. Bot. 51, 1471-1480 (1973).

101. Tyree, M.T., Hammel, H.T.: The measurement of the turgor pressure and the water relations of plants by the pressure-bomb technique. J. Exp. Bot. 23, 267-282 (1972).

102. Ussing, H.H.: Some aspects of the application of tracers in permeability studies. Adv. Enzymol. 13:21 (1952).

103. Van't Hoff, J.H.: Une propriété général de la matière diluée. Svenska Vet. Akad. Handl. 21 (17, 43) (1886a)

104. Van't Hoff, J.H.: Lois de l'équilibre chimique dans l'état dilus gazeux ou dissous. Svensk. Vet. Akad. Handl. 21 (17, 217) (1886b).

105. Van't Hoff, J.H.: Die Rolle des osmotischen Druckes in der Analogie zwischen Lösungen und Gasen. Z. Phys. Chem. 1, 481-508 (1887).

106. Van't Hoff, J.H.: Zur Theorie der Lösungen. Z. Phys. Chem. 9, 477 (1892).

107. Waley-el dine Sameh: Daily Life in Ancient Egypt. München: Verlag George D.W. Callwey (1964).

108. Walter, H., Steiner, M.: Die Ökologie der ostafrikanischen Mangroven. Z. Bot. 30, 65-193 (1936).

109. Yayanos, A.: Equation of state for P-V isotherms of water and NaCl solutions. J. Appl. Phys. 41, 2259-226 (1970).

# Subject Index

**Molecular Biology, Biochemistry and Biophysics**
Editors. A. Kleinzeller, G.F. Springer, H.G. Wittmann

Vol. 1: J.H. van't Hoff, Imagination in Science. Translated into English with notes and a general introduction by G.F. Springer. 1 portrait. VI, 18 pp. 1967
Vol. 2: K. Freudenberg, A.C. Neish, Constitution and Biosynthesis of Lignin. 10 figs. IX, 129 pp. 1968
Vol. 3: T. Robinson, The Biochemistry of Alkaloids. 37 figs. X, 149 pp. 1968
Vol. 5: B. Jirgensons, Optical Activity of Proteins and Other Macromolecules. 2nd revised and enlarged edition. 71 figs. IX, 199 pp. 1973
Vol. 6: F. Egami, K. Nakamura, Microbial Ribonucleases. 5 figs. IX, 90 pp. 1969
Vol. 8: Protein Sequence Determination. A Sourcebook of Methods and Techniques. Edited by S.B. Needleman. 2nd revised and enlarged edition. 80 figs. XVIII, 393 pp. 1975
Vol. 9: R. Grubb, The Genetic Markers of Human Immunoglobulins. 8 figs. XII, 152 pp. 1970
Vol. 10: R.J. Lukens, Chemistry of Fungicidal Action. 8 figs. XIII, 136 pp. 1971
Vol. 11: P. Reeves, The Bacteriocins, 9 figs. XI, 142 pp. 1972
Vol. 12: T. Ando, M. Yamasaki, K. Suzuki, Protamines. Isolation, Characterization, Structure and Function. 24 figs. 17 tables. IX, 114 pp. 1973
Vol. 13: P. Jollès, A. Paraf, Chemical and Biological Basis of Adjuvants. 24 figs. 41 tables. VIII, 153 pp. 1973
Vol. 14: Micromethods in Molecular Biology. Edited by V. Neuhoff. 275 figs. (2 in color). 23 tables. XV, 428 pp. 1973
Vol. 15: M. Weissbluth, Hemoglobin. Cooperativity and Electronic Properties. 50 figs. VIII, 175 pp. 1974
Vol. 16: S. Shulman, Tissue Specificity and Autoimmunity. 32 figs. XI, 196 pp. 1974
Vol. 17: Y.A. Vinnikov, Sensory Reception. Cytology, Molecular Mechanisms and Evolution. Translated from the Russian by W.L. Gray and B.M. Crook. 124 figs. (173 separate ill.) IX, 392 pp. 1974
Vol. 18: H. Kersten, W. Kersten, Inhibitors of Nucleic Acid Synthesis. Biophysical and Biochemical Aspects. 73 figs. IX, 184 pp. 1974
Vol. 19: M.B. Mathews, Connective Tissue. Macromolecular Structure and Evolution. 31 figs. XII, 318 pp. 1975
Vol. 20: M.A. Lauffer, Entropy-Driven Processes in Biology. Polymerization of Tobacco, Mosaic Virus Protein and Similar Reactions. 90 figs. X, 264 pp. 1975
Vol. 21: R.C. Burns, R.W.F. Hardy, Nitrogen Fixation in Bacteria and Higher Plants. 27 figs. X, 189 pp. 1975
Vol. 22: H.J. Fromm, Initial Rate Enzyme Kinetics. 88 figs. 19 tables. X, 321 pp. 1975

Distribution rights for U.K., Commonwealth and the Traditional British Market (excluding Canada): Chapman & Hall Ltd., London

**Springer-Verlag**
**Berlin Heidelberg New York**

**Encyclopedia of Plant Physiology, New Series**

Edited by A. Pirson and M. Zimmermann

The well-known and successful Encyclopedia of Plant Physiology, conceived three decades ago by W. Ruhland, was concluded in 1967 with the publication of the last of the 18 volumes. The Encyclopedia is still an important reference work useful for both research and teaching.

But plant physiology has continued to develop and considerable advances have been made since the publication of the Encyclopedia. Biochemical and biophysical methods, and the methods of modern molecular biology in particular, continue to stimulate research.

Extensive discussions between Springer-Verlag and numerous scientists resulted in the concept of the Encyclopedia of Plant Physiology, New Series, - under the editorship of A. Pirson, Göttingen, and M. Zimmermann, Harvard. The volumes of the New Series are dedicated to special topics, they will be smaller and less expensive, and thus will appear within a shorter publication time. For more extensive areas multiple but largely self-contained volumes are planned, and editors of related volumes will maintain close contact with each other. The New Series will be written entirely in English and each volume will be edited by editors who are experts in the fields concerned. The editors and the publisher hope that the Encyclopedia of Plant Physiology, New Series - will in time build up a comprehensive, modern treatise covering the whole field of plant physiology. Such a work will stimulate collaboration among specialists of different disciplines, offer detailed information for teaching at university level, and enable graduate students to find their way into the current research areas of any field.

Vol. 2: Transport in Plants 2. Edited by U. Lüttge and M.G. Pitman. With a Foreword by R.A. Robertson. With contributions by numerous experts. Part A: Cells. Part B: Tissues and Organs.

Vol. 3: Transport in Plants 3. Intracellular Transport and Exchange Mechanisms. Edited by U. Heber and C.R. Stocking.

Vol. 4: Physiological Plant Pathology. Edited by R. Heitefuss and P.H. Williams. With contributions by numerous experts.

**Springer-Verlag**

**Berlin Heidelberg New York**